Off the Grid Living

How You Can Live Off the Land and Become Self-Sufficient through Homesteading and a Backyard Guide to Raised Bed Gardening

Contents

Part 1: Living off The Grid

A Guide on How to Live Off the Land and Become Self-Sufficient Through Homesteading

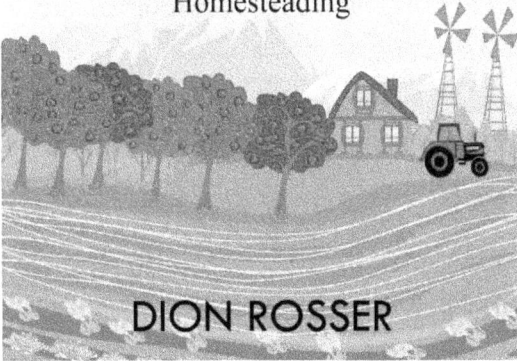

LIVING
—OFF THE—
GRID

A Guide on How to Live Off the Land
and Become Self-Sufficient Through
Homesteading

DION ROSSER

Introduction

What is living off the grid? Take a minute to think about your definition. Write down the things you've heard about living off the grid and keep it handy but set it aside for a moment because you will get back to your description in a few minutes.

You bought this book because it is up to date, easy to understand, and great for beginners. It contains hands-on methods and instructions to start you living off-grid or as much off-grid as you wish.

Take the time to decide what you think it means to be off-grid now; you might just get a pleasant surprise or two in these pages.

You will learn what the steps to live off-grid are.

You'll gain an understanding of what it takes, the pros and cons.

When finished reading, you will know:

1. What you need to get started.

2. How to create a homestead layout.

3. The style of dwelling you want.

4. The water and sewer choices.

5. How to heat and use energy.

6. Gardening techniques.

7. How to raise livestock (and why you want to start with chickens!)

8. Ensuring you preserve the food.

9. How to make money from your home.

Numerous topics go into living on a homestead and not depending on anyone else for your survival. Whether you take up hunting or live a more pacifist life by eating only vegetables and fruits, this book can make sure it happens.

Now, with the world in upheaval, hundreds of thousands of deaths from COVID-19, and potential issues of buying enough food from a supermarket, it's time to learn how you can have a sustainable life off-grid. You'll even gain skills that you can make money from when you are ready to employ them.

Chapter 1: What is Living Off the Grid?

There are as many definitions of off-grid living as there are resources available online in the form of homesteading, do-it-yourself, prepping, and green living websites, each differing in the technicalities, but there's one constant in all of those definitions—it's a lifestyle in which you don't rely on everyday utilities such as electricity, gas, and water.

Let's look at the definitions to get a better idea of what living like this entails.

Living off-grid is a lifestyle and system that is independent of remote infrastructures such as gas and electricity. It mentions that all off-grid houses are autonomous, as they do not depend or form on sewers, municipal water, and other utilities.

MacMillan Dictionary defines it as avoiding the commodities, including water, electricity, and gas.

Urban Dictionary (yes, I went there) states something a bit more precise describing it as "Living in an unrecorded and untraceable area through regular means." If you want an example, Ron Swanson from Parks and Recreation is a good one. Urban Dictionary further jests this way of living is a term for a person who uses no form of social media. Yeah, right!

Seeing the definitions, we can theorize that living off-grid refers to a lifestyle with a house completely free from public utilities and employs sustainable living by using renewable energy.

Let's now consider all these points one by one.

Being Independent of Public Utilities

What are the public utilities you can think of off the top of your head? Municipal water, sewers, and natural gas are what we're talking about.

Being independent of municipal water means you aren't connected to the town water and have your own off-grid water system instead. We will discuss more on that later.

Being independent of sewage involves having your field bed, a septic tank, an outhouse, and a propane/composting off-grid washroom.

Last, being free of natural gas saves you a ton in terms of bills, which leaves you money you can spend on having your wood stove, firewood, and other necessities you can use to cook food and heat your house with.

If being completely self-reliant and independent sounds too good to be true, and if it gets you excited, then choosing this lifestyle can be the right choice for you. Not just yet, you must take baby steps. The first question to ask yourself is how off the grid do you want to be, and if you're a person with a family, what do they think of this? This style of living doesn't appeal to everyone.

Being Independent of Electricity

This involves you being independent of an electrical power grid, which you can accomplish by either living without electricity or by developing your systems, such as using solar power, a generator or even turbines that run on wind. That's where the term "off-grid" originates; *the grid is the electrical network.*

Levels of Living Off-Grid

The amount of self-sufficiency you want to integrate into your life is yours to decide. This means different things to different people. There are varying levels of self-sufficiency. Certain people believe that you aren't following this way of life if you still must rely on propane and diesel as fuel for your generator or for heating your house. But others believe that they are merely by installing solar panels in their house in the suburbs. The latter can even sell their generated power to the local network.

Are You Still Off-Grid if You're Still Using Fuel?

Yes. As a hypothetical, let's consider that you're living in a cabin in the woods, heated by firewood, have an outhouse for toiletry, and have your water source. That counts as off the grid. You can use solar power and wind power to generate electricity, in which case you won't be using diesel, oil, or propane. That's considered being completely off-grid, but if you use fuel, that's you being off-grid *only to a certain degree.* Remember, we talked about varying degrees of this lifestyle. The definition is up to you to produce.

Using Solar Power Isn't the Criteria

Technically, if you have a house in the suburbs and are using solar power as your primary source of electricity, you're still connected to the electrical grid. So, doesn't that beat the purpose of living off-grid? In most American states and Canadian provinces, you have the option of selling your solar power, a process that's known as *net metering.*

Are You Allowed to Use the Internet?

Totally. Where you're getting good cell service, you can use the data on your phone to access the Internet. You can have a wireless hub, which allows you to stay connected even if you're residing in a very remote location. Internet access is unbelievably valuable to folk who want to live this way, but still want to generate money by, say, freelancing to sustain financially. And you can use satellite Internet or sometimes visit the nearest town to access public Wi-Fi such as a coffee shop or a fast-food joint. Ask yourself what you want to do with the Internet access and then plan accordingly.

What Does Live Off-Grid Mean to You?

The definition—after including the barebones—is up to you to decide. Early in the process, decide how, what, why and where you plan to get off-grid so you can make plans and strategies to accomplish your goal. For many people, living off-grid is a dream come true, and for others, it's a way to escape the hustle-bustle of living a hectic life in the city. Do you wish to be more self-sufficient? Are you yearning for independence from the power grid? Are you doing it to enjoy nature more? Do you intend to pursue this because it suits your personality, and you consider it quite a lucrative lifestyle? Make up a list of points as to why you crave to do it and what's in it for you. For example, what are the pros and cons of this lifestyle, how rewarding is it, and how tough can it be, what should you do if an emergency occurs, and last, are you cut out for this?

Why Do You Want to Live Off-Grid?

Everyone has their reasons, and I'm sure that you have them too. With those in mind, let's look extensively at the general why's and see if they resonate with your reasons.

1. It makes you self-sufficient.

2. This form of living is sustainable.

3. You'll be relying on renewable energy.

4. This is a more environmentally responsible way to live life.

5. Off-grid living is more practical since we're reusing and recycling and managing our resources.

6. Since you're using fewer resources, you'll leave be leaving less waste.

7. This lifestyle means you get to admire and appreciate nature in a way that's simply impossible in the city. You're more attuned to nature, which will improve your lifestyle, making it healthier.

8. You'll be actively doing chores around the house, working, and striving to sustain yourself, which will make you healthier. The physical benefits are way too many to count.

9. It's a rewarding lifestyle.

10. It will make you happy and content because of more activity and less stress.

11. It will attune you to your roots. Our ancestors used to live this way thousands of years ago. By connecting with your roots, enjoying nature, and utilizing natural resources, you'll find internal satisfaction.

12. You'll be passing that knowledge to the next generation: your kids, students, and younger people. You can share your journey online on YouTube or in a blog, sharing your experiences with people online, and as we mentioned earlier, yes, you can use the Internet and still be off-grid.

13. It is an excellent way to avoid consumerism.

14. You'll be leading other people by your example. Maybe they'll follow and live off-grid after they see the positive impact it has had on your life.

15. The food you'll harvest and prepare will be healthier, as there will be no pesticides or other chemicals used in growing it.

16. The air out in nature is fresher and more exhilarating than the city's air.

17. You must build stuff for yourself including tables, chairs, stoves, makeshift futons, and that sort of thing. This will not only add to your carpentry skills, but will also make you more active, because making things yourself will be a strenuous task.

18. Independence. You'll be more independent of everything.

19. It shall make you more financially responsible.

20. It's a much more socially responsible way to live life. Plus, the lifestyle is safer as compared to living in a city.

21. When you have your homestead, you'll be creating resources for yourself and preparing your own food. Once you make more resources than you need, you'll sell them or sharing them with your local community, which will not only help you earn money but also contribute to your local neighborhood. You get to give back. You can even help local restaurants and eateries by selling them your organic vegetables and fruits. If you're more inclined toward charities, you can contribute to the local food bank.

22. Your backyard will be the entire forest, hillside, and wilderness, which allows you to take long walks, hikes, treks, and explore your surroundings. You'll have more freedom to keep pets. How about dogs? Imagine trekking with them close at your heels.

23. Out in nature, it's peaceful, quiet, and calm.

24. As an off-gridder, you'll always have something to do, whether it is recreational, or something related to your work. You'll mostly be busy, which is an excellent way to pass the time and improve yourself.

25. Living off-grid is taking a long vacation without leaving your home. Just think about all that nature, all that flora, fauna, green trees, placid lakes, and mysterious paths leading into the untamed wilderness. Doesn't that just fill your heart with excitement?

What's In It for You?

1. You'll Get to Escape the System

Even though it's a bit of a political statement to move out into the wilderness and to make you aloof from the "system," once you are out there, you'll feel freer because you'll be rid of debt slavery, corruption, greed, and materialism. If it sounds a little too good to be true, it's because it is. You'll be getting a sense of independence once you aren't shackled by quite the oppressive and intrusive system.

2. You Will Learn Preparedness and Survival Skills

This way of life makes you prepared for many things going awry, especially if you take Murphy's Law into account. You'll be prepared for the worst, honing your survival skills as you do so. This is called prepping.

3. You'll Get to Live a Sustainable Life

You'll reduce the number of resources you consume, thereby making your life self-sufficient and sustainable. You'll be consuming what you produce and will only have room for dire necessities. No excess. Your life will be more balanced.

Two Examples of Off-Grid Living

1. The Cheap Way

Others call it roughing it; I call it the cheap way. It's because this method is not only cheap but also allows you to be completely off-grid. You'll build a small house for yourself, such as a dry cabin. There's no water or electricity in that house; it's essentially just a structure. You'll have to build an outhouse and create a garden for your food, the latter of which is called homesteading. Since this is a cheaper option, know that you'll be relying on one generator for your power. Most days, you'll spend time by the fireplace, the stove, and you will be using candles for light. You'll be expected to get your water, as you won't be near a well or a source. You can either collect rainwater or go into town now and then to buy water. You can also build an underground bladder to store water, but this will cost you more money than this example allows. This form of living is called the Alaskan style. You will be washing all your clothes by hand. Going to the laundromat is not recommended, as it is counterintuitive to your lifestyle. Another thing that you must consider in this style is how to take a bath since you have no running water. You're not going to have an electric stove, so you'll be using more primitive alternatives, and the same goes for refrigerating your food. If you're considering this form of living, consider the pros and cons first, which we shall discuss later, first in an overview, then in a detailed manner.

2. Half and Half

Also known as small-scale homesteading, the half and half style has you depending on the system, but not completely so. You can stay on the grid in this style. Half of your life will rely on the system (power, water, gas, sewage), and the other half will involve you homesteading while trying your best practicing this lifestyle. You'll still have access to the town you're living in. For instance, you'll get to cook your food on an electric stove, wash your clothes in a washing machine, etc. This form of being off-grid is recommended for those of you just starting. Consider it as tapering off the grid, one step at a time until you're ready to dive in completely.

The advantages and disadvantages of living off-grid:

Advantages

The biggest one is that you are self-sufficient. Consider this: the power goes off in town or your municipality. Thankfully, you won't be affected. The same goes for any emergency that takes place at, say, the water plant, for example.

You won't be spending exorbitant amounts of money on your bills.

Last, freedom is the biggest advantage, and you can't put a price on that.

Disadvantages

You must move a lot because most places won't allow you to be off-grid. Your resources will be scarce if you move to a place that's not close to town.

You must consider the impact it will have on your family.

It could mean legal trouble. Before you try this way of living, consider the local laws.

Your life will be drastically affected, such as your inability to run a shower, iron your clothes, and wash clothes and utensils.

The Rewards and Challenges

1. You must source your food. Your priority will be to secure your food sources. This entails getting pesticide-free produce, establishing a secure source, and arranging meat that's humanly raised. This will require a great deal of challenging work on your end because you must set up the

whole system and maintain it, harvest food, and butcher the animals. Last, once you have all the raw materials, you must be careful that they don't go to waste, but remember that this will pay off monumentally.

2. You will have to build your own home. Before delving into this venture, you must brush up on your home building skills and make sure that everything you build is durable. In return, you will get complete freedom for choosing the materials, and you can always get non-toxic stuff and renewable resources for your home, thus positively contributing to the environment.

3. Nothing gets done if you don't do it, so if you're the type of person who lives a procrastinating and passive lifestyle, this honestly might not be the right fit for you. You must take into consideration that you will have to chop your wood, grow your food, and empty your toilet. What do you get in return? Freedom, yet again. You're not going to have to report to a superior.

4. You won't be at the mercy of the grid, which will allow you to weather all the storms that come your way. You're going to have to work a lot, sure, but you'll also get consistency, stability, and hardiness in your character.

5. People will not get you. They'll think of you as eccentric, viewing your lifestyle with apprehension, curiosity, and skepticism. It will also be a conversational hurdle, as you won't be able to relate to people, nor will they be able to relate to you.

How to Get Started Living Off the Grid

1. Get a Wood Stove

Installing a wood stove in your home will heat your water, your food, and allow you to dry your clothes even when it's raining. It will also save you a fortune in terms of energy bills.

2. Learn From the Pros

There are a ton of guides (including the one you're reading) online that can help you get motivated on how to get started.

3. Learn How to do Things by Hand

This is key. You'll be doing most of the work by hand since there won't be all that many convenient devices at your disposal. A few examples include kneading the dough, using a manual coffee grinder, and using a handsaw.

4. Prioritize Your Evenings

Now that you've decided to try this lifestyle, understand that your evenings won't be spent in front of the computer or the television. You can use that free time to learn skills such as cooking, sewing, whittling, and knitting.

5. Pick Out a Location

If you're serious about going off-grid, you should start looking for a piece of land where you can build your homestead. It's not an easy task to find affordable acreage far away from the city that has all the commodities and amenities nearby if you should need them. It's going to require thoroughly researching real estate, land prices, and location.

Chapter 2: Are You Cut Out for Off-Grid Living?

Are You Aware of the Challenges of Living Off the Grid?

Because let's be honest, as popular and appealing off-grid living is, it comes with its own sets of challenges and considerations. Know beforehand if this is something you're cut out for or not. Before jumping right into this lifestyle, let's consider a few of the challenges in detail.

An important statistic to consider here is that upwards of 200,000 Americans now live off-grid.

There's a steep learning curve to this way of living, but once you're past that curve, things will start to fall in rhythm.

1. The Location

Location does not just pertain to people who live in the city. As someone working to live an off-grid life, you must consider that the land you pick is just about rural enough that there are no utilities available, such as electricity and gas. Now, second, consider the land that you need for your homestead. What are the essentials you need to sustain yourself? Are you planning on raising livestock? If so, you must get land that offers shelter, stores their food, and allows them to graze. You will also need a water source such as a well. Last, you'll need space and land for your waste management, recycling system, and your garden.

There's also the issue that many municipalities don't allow you to dwell as an off-gridder, requiring that you hook your home up with the power grid. In certain cases (in a few cities), they'll allow you to go off-grid with the added caveat you supplement your power with alternatives and don't cut off completely. All this calls for buying land that's nowhere near a city or a municipality.

2. The Power Source

The next important bit to consider is the power source. Yes, you can go into *ye olde* times by cutting yourself off completely and not relying on any power sources, but if you do decide to have a power source, it'll make your life a bit easier. The easiest route here is getting solar panels. It's a long-term investment with a decent ROI (return on investment). Another option is a generator. There are a dozen alternative sources at your disposal, and they all depend on what's your budget, what's your location, and how much power you need. For a starting point, check your electricity bill and see how much energy you are using. Now, plan accordingly.

Once you've set up your power, you will quickly realize that you are consuming more energy than needed. Consider this as a withdrawal symptom from your days living on the grid. Conserving energy will be beneficial here. Try running only one appliance at a time and do that when it's daytime so you can get the most out of your power source if it is solar energy. Don't rely all that much on your energy source, though. You are, going off-grid. Ease yourself in with alternative methods, such as using a stove to heat your food and home and using gas instead of an electric stove.

3. The Food

As an off-gridder, you'll be sourcing your food. This means growing everything you need to eat, raising cattle and livestock, and getting a garden established. The kitchen garden you establish in your homestead must expand gradually, as you will start producing more food than you need so you can sell it and generate income to supplement your lifestyle.

Possibly, the biggest issue you'll face in terms of food is a balanced diet. You need to have variety in your food since it's just impossible for everyone to raise livestock at the start of their homesteading. So, what do they rely on in the initial phase? Veggies. That's right. You can produce vegetables in your garden.

Another challenge that you'll be facing is cooking and preserving food. Both these skills are essential in homesteading and require diligence for the offseason in which your garden isn't producing as much.

Many homesteaders go the hunting route. If you're the sort who likes to hunt, this will be an excellent opportunity for you to hunt game, skin them using a tactical knife, butcher them, prepare them, eat them, and preserve them.

4. Water Supply

You must have a water supply nearby. It should be a reliable water supply and one you can depend on when things go awry. Trust me, things can go awry fast when living on your own in a rural setting with nothing to rely on but yourself. Add to that the unavailability of water, and you'll have a complex situation with no harvest, no cattle, and nothing to sustain you. Some lands are located near freshwater sources such as rivers, but they are rare to find.

Instead, bore a well and use a hand pump to draw the water out whenever needed.

Another important thing to consider when making the shift from your "normal" lifestyle to an off-grid lifestyle is that you shall miss running water. It's a commodity that we take for granted, and when it's not available to us, such as in the case of being off the grid, we'll feel out of place. A workaround includes getting a generator so you have running water at hand, but one factor about generators is that you need to keep buying diesel for them. Is this something you see yourself doing, or would you rather save costs on fuel?

Last, you must keep the legal aspect of it all in mind. Are you allowed to use the water on the land that you've bought? Usually you can use the water in your property for your personal use with no issue, but, say, if you are using water for livestock and farming, keep in mind the laws of the municipality. Consult a lawyer beforehand, if necessary.

5. Time Management

Think of living this way as having multiple jobs all simultaneously. From harvesting food to cleaning the outhouse, tasks are abundant at hand. You're going to have to manage your time expertly so that everything goes forward smoothly. There's no room for procrastination here. If you're sowing seeds, they should be sown in the correct season. If you have animals, you must feed them regularly and milk them on schedule. To protect your homestead, you must set up fences. To store your harvest and food, you must build shelters.

All that work will keep you more than occupied and on your toes. It gets tiring and gets jarring at times too, but overall, once you've picked up a rhythm, you'll thank your past self for it. What's the secret to it all? Time management. You can keep a journal or a to-do list with you. If you still have a phone, you can store your schedule in that.

To find time to do all the above-mentioned stuff and more, you must become an early riser. Also plan out the next few months or a year for long-term projects. Besides keeping track of your time, also keep track of the weather patterns.

6. Budget

Let's get one thing straight—this is not a cheaper option. That false image is the romanticized stuff fit for movies and books, where the protagonist moves out of the burbs and heads into the wilderness where he finds internal peace. The last part is not exaggerated. You'll get your fair share of internal peace, but it will come at a cost. There will be a slew of upfront investments that you must make to succeed. Also, you shall have to keep up with the present rates for livestock and their feed, and the cost of growing your crops.

The most important thing where a considerable amount of your budget will go is the power source. Setting it up will require you to make a huge investment in solar panels or generators. If we're realistic here, this will cost you tens of thousands of dollars, especially considering that you will have to set up between 15 to 30 solar panels to support the needs of your homestead.

If that sounds like too huge of investment, start with baby steps. First, grow your food and cooking it, then move on to reducing your carbon footprint, and then become more conscious of your power consumption. By taking these measures, you'll end up not only saving money for your new lifestyle but also get in the essential habits needed to survive.

7. Isolation

Let's consider the hypothetical protagonist who found inner peace by heading off into the wilderness and cutting all ties with society. That's not always going to be the case. A lack of human interaction will serve as a challenge to you if you're a social and extroverted person. Yes, most homesteaders do start their new lifestyle to escape from the city and its hectic life, but it need not be a solo venture. Contact other like-minded people and form an off-grid community where you can go whenever you feel alone and isolated.

If there isn't a community nearby you don't fret. There are dozens of communities online where veteran off-gridders share their advice with newbies, and people share their tips, tricks and experiences. Consider joining them.

Are You Ready for an Off-Grid Lifestyle?

1. Your Mindset

You should have a conservation mindset to pursue this lifestyle. In our regular homes, we become accustomed to a lifestyle of excess. Running water is available whenever we want it for washing, laundry, and taking baths. There's always electricity to rely on for our gadgets, for changing the temperature of our rooms, and for running practically everything ranging from stoves to garage doors. You and I pay our bills every month, never once considering where all our resources are going and where they come from.

There's a need for a radical shift in your mindset when you decide to go off-grid, especially in terms of resources because they are limited and deplete the more you rely on them. How does one avoid that? It's simple:

> • At a given time, use only one appliance. Say, at breakfast, you need to use the blender, toaster, and the coffeemaker; you will have to use them one at a time instead of using them simultaneously.

> • After you have used a charger, unplug it.

> • Use power-heavy appliances such as washing machines and dishwashers in the daytime, particularly if you're using solar power.

• Take baths only in the afternoon when the water is at its hottest because of the sun.

• If you can cook with propane, all the better.

• Your priority for heating your house should be a woodstove rather than any electrical heater. There are no thermostats out here.

• Open windows as often as you can for ventilation and nighttime cooling.

2. A Design Specific to the Site

The architecture of the house is critical when considering going off-grid, or even just trying to be super-efficient in your regular home.

Remember, you must consider the building's passive solar design—as in placing windows and skylights to maximize sunlight or exposure to the southern side—and the orientation of the building. You'll be getting free heat and light out of the deal if you plan your house accordingly.

Think about having a super-tight enveloping building to help you manage energy loss, particularly when there's an enormous difference between the external and internal temperatures.

Everything that you do should be energy conservation centric.

3. Power Generation

An average American home utilizes around 10,000 kilowatt-hours per year. Before jumping the gun, first, analyze your bills for a pattern and see how many kilowatt-hours you are utilizing on a monthly and annual basis. Then consult with your architect and determine the shift in your lifestyle and how that can improve your power costs.

Think about all the appliances you use every day, such as your irons, dehumidifiers, hairdryers, and hair straighteners. Have you considered how deeply how you are tied to the grid and how invasive it is in your everyday life?

How are you planning to power your homestead? Will you be using solar power, geothermal, or wind? Regardless of the source, you need to have backup power in the form of a generator.

4. Water Collection

In terms of essentials, a renewable water supply is important. A common way to go about it involves a well or using freshwater. After you have established a water source, pick out a plan as to how you will get it to your home. Will you use pumps and a filtration system?

Once that water is at your home, you'll need a holding tank for it. If you're in an area that gets heavy rainfall, you can set up rain barrels to supplement your water consumption. This is gray water, not recommended for drinking use. Instead, use it for your landscaping needs.

5. Waste Disposal

This bit is a little complicated as you're both bound and unbound from the government simultaneously. Yes, you are still (hypothetically) off-grid, and normal building codes are not applied here, but there are entities such as the Environmental Protection Agency and other health departments that require you to dispose of your waste safely according to their regulations.

One way to go about it is by installing your septic and drain-field system that will return all waste into the ground, where it will all be utilized in nutrients. Yes, the system will take up a lot of space when you're installing it, but once it is all done and dusted, it will disappear. Last, you will need to maintain both these systems.

Is Living Off the Grid the Right Choice for You?

Understand that this way of living isn't for everyone. Not everyone is cut out to tough it out in the wilderness. Although, if you're feeling overwhelmed and thinking to yourself that you might be ready for this, here are a bunch of questions that shall put it into perspective.

1. Do You Value Your Time?

When you're on the grid, you're tied down by a system that constitutes a job, giving time to your television and computer, going to and from a grocery store, managing your social life on top of your work life, and seeing to all the expenditures that come with the territory of living in a city. To say this lifestyle consumes a lot of your time would not be an overstatement.

Once you're living off-grid, most of the time you get free from your homesteading tasks will be yours. You can even quit your job and earn money by selling your organic produce. You're not going to spend time in front of the computer, you won't have to keep on checking your phone, and you won't be sitting passively in front of the television. All that free time will be yours to do with as you please. You will be working on your schedule and for your benefit. As grueling as the lifestyle is, it is just as rewarding.

2. Your Friends and Family's Reaction?

This will be a fantastic opportunity to weed out people who aren't your real friends. When you tell them about your decision to live this way, those who care for you'll support your choice and tell you to go for it, whereas those who don't have the best intentions for you in their heart will tell you many reasons you should not do this. A few of your friends and family members will be inspired by your example, others will be jealous, and the rest will be indifferent. It's up to you and you alone to decide to be off-grid or not.

3. Will You Have to Quit Your Job?

Once again, it is up to you. If you are planning on being a full-time off-gridder, then, yes, quit your job and focus solely on your new lifestyle. If, on the other hand, your job is essential, then consider the half-and-half method we discussed, in which you'll be off-grid but stay in the city so you can commute to and from your job. Then it is up to you to decide. If you have a family, then consider their wants and needs, ask for their opinion, and then produce a system that can sustain you.

4. Can You Afford It?

Start saving money the moment you go off-grid. You don't have to have all the money right away. You can do this in easily manageable batches that won't be heavy on your wallet.

5. What to do with the Money You Save

The saved money will go to all the costs such as energy, water, waste management, and food.

6. Do You Know What You're Doing?

Kudos to you if you know what you're doing, but if you're unsure about what steps to take, read up on a couple of resources, consult online, and watch as many videos on YouTube as you can. Asking veteran off-gridders for advice is another way of making sure that you're doing the right thing.

7. Do You Have Contingencies?

Again, with Murphy's Law being ever-present, all that can go wrong will go wrong, so you need to be mentally – and resourcefully – ready for that. Write it in your journal. Keep track of all the things that can go south and produce a series of steps to tackle them.

Chapter 3: The Pros and Cons of Off-Grid Living

We discussed the advantages and disadvantages briefly in the previous chapter. Let's take a more detailed look at them now.

Pros

1.Peace and Quiet

It's utterly quiet and calm when you start living off-grid. Gone are the sounds of traffic horns and people cursing at each other in traffic jams. You are trading in the overwhelming sound stimuli for a feeling of peace and quiet that will leave you feeling peaceful. You won't have a constant bombardment of noise after noise, plus all the other forms of pollution that add stress and anxiety to your life, such as all the many screens you are used to watching—phone screens, TV screens, and LED advertising panels. Because of this lack of incessant stress and anxiety, you will be less overwhelmed in life. You'll be getting a digital detox when you move off-grid. At first, you will have your fair share of withdrawal symptoms, but once you're attuned to your new lifestyle, you shall appreciate it more for all the perks it grants you.

2.Quantum of Solace

If you are planning to go off-grid all by yourself, it will minimize your lifestyle in a whole new way. You won't be interacting with all the people that were always around you. This includes relatives, friends, colleagues, and neighbors. You shall be living on your own, having to answer to no one. Initially, you shall feel a bit out of place, sure, but as you get into your new lifestyle, you shall appreciate cutting off all (well, most) ties with the world. Your life will have fewer distractions in it. It will be a less intrusive life, one where everyone won't be bothering you all the time, which translates to less stress.

3.Land Security

Doesn't every one of us want to own our place, a place that we can call home? Well, here's your opportunity to do so. Not only is the property in rural areas cheaper, but there is also a lot of it. Once you buy your land, you will not have to worry about things like rent and mortgages. You'll be in total control and ownership of your piece of land and can do whatever you want to do there, such as building new extensions, keeping livestock, expanding your homestead, setting up your garden, and so on.

4.Surrounded by Nature

This is probably the best advantage of moving off-grid. Greenery and nature are scarce in cityscapes, limited to parks and small gardens and the odd tree here and there. When you adopt an off-grid lifestyle, you find yourself pleasantly surprised by all the surrounding nature such as: green grass, lush trees, babbling brooks, peaceful lakes you can swim or boat in, and the clear blue sky unadulterated by the fumes of vehicles and factories. You'll become healthier in just one month in your new lifestyle and will take long walks to appreciate nature more. This will get you in touch with your roots. It will also make you less anxious and depressed. All this nature around you will be very therapeutic and inspirational.

5.Time

Time tends to stand still in the wilderness. Granted, you will be busy managing all the aspects of your new lifestyle, but all the time you get free from that will be yours and yours alone to do with as you please. There will be no interruptions. You will get to decide your schedule for the day, answering to no one. You can spend this time—as we discussed earlier—in learning new skills and developing new hobbies. You'll be able to do whatever you want, given you go full-on off-grid. Doesn't that sound swell?

6.Opportunity for Introspection

Once off-grid, you shall have an abundance of time and energy to explore all the ideas in your head, once and for all. This includes everything from your daily thoughts to plans about starting a new hobby or finally getting that book you promised yourself you'd write. Learning a new skill goes hand in hand with off-gridding. You'll be able to introspect and discover what sort of person you are, what you inherently like and dislike, and where do you want your life to go from this point forward. All this introspection will grant you more insight into your life. And the best part about it is that you have time to meditate.

7.Fresh Air

Do you know how important it is to breathe fresh air? If not, then consider all the thousands of vehicles and factories that contribute to the pollution inside the city. Essentially, what you're doing is slowly suffocating yourself in the city. Once out in the open surrounded by fresh air, you shall feel more energetic, cleansed, and healthier. When it's nighttime, you can see all the stars, and when it's daytime, your view of the sky will be pristine.

8.Mindful Living

Meditation and mindfulness are critical to personal development and living a healthier lifestyle. Once you're out in the open, you can set up a schedule for when, where and how long to meditate.

Declutter your mind. This will catalyze the simplicity in your life since all the choices that you will make from this point forward will be deliberate and yours to make.

9.Limited Access

Living off-grid will limit your access to the closest town and technology, making you more self-reliant and less dependent on systems and tech. All this adds to more time on your hands. This is only if you decide to go completely off-grid.

10.Social Life

If you are an introvert and living a social lifestyle drains you of your energy, the disconnection from people and social responsibilities will allow you to live life on your terms without the pressure of meeting and staying in touch with people.

Cons

1. Start-Up Costs

Everything involving off-gridding comes at a price. Unfortunately, that price is a bit too steep for most people. This is sometimes the only factor that deters most people from making the shift. Purchasing all the equipment, getting it installed, and maintaining it is awfully expensive. You're going to have to start saving up at least a year or two before you can go completely off-grid. It isn't just the start-up costs you must consider; there are dozens of factors at play here that will all require you to have deep pockets. But this should not be a deal-breaker for you, since you can start off-gridding in batches, breaking it all down into smaller modules that are less taxing on your money. Remember to take one step at a time.

2. Maintaining Everything

If you're a one-man (or woman) off-gridder, you'll find yourself busy with maintaining everything round the clock. As we mentioned before, nothing gets done if you don't do it. Whether it's renovations around the house or repairing the worn-out fences surrounding your property, maintaining everything will be not only tiring but also expensive. There will be days when you question whether it is worth it. You'll have to remind yourself the reason why you started off-grid living in the first place, and that it is indeed worth it.

3. Waste of Energy

All the energy you won't be utilizing will go to waste, and here's the thing—you will not be conscious of your energy consumption all the time. Think of it like this: are you conscious of your energy consumption at home? It takes more than a couple of months to get into the conservation mind-set where you begin to be more conscious of your resources, but for the first few months, you will end up wasting all the energy you won't need. This is so because surplus energy cannot be stored anywhere. A workaround to that is using only those necessities when you need them, then unplugging them.

4. Difficulty Communicating

As a full-time off-gridder, you'll cannot stay as connected with people from your previous life. If you intend to go completely off-grid—as in no computer, no phone, and no Internet—the prospect may sound appealing at first, but once you are doing it, you will realize that it's difficult to not communicate with people. Also, if emergencies occurs, where will you go, or to whom will you turn? You cannot call people when you want and vice versa. Honestly, this is the biggest downside of living off-grid. You'll be disconnected from your friends and family and only be able to meet them when you make one of your trips in town to get supplies and equipment. Are you ready for that?

5. Disconnect From Technology

There are degrees to which you decide to live off-grid. In many cases, you will not disconnect from technology and will have your phone and television and computer at your home, but where you're going full-blown off-grid, the disconnect will come at its costs. You won't be able to keep up with the news, won't be able to work remotely, and won't be in touch with your friends and family. This is why most off-gridders don't go completely tech-free.

6. Limited Social Activity

Since you're living off-grid, you won't be able to attend parties, social gatherings, and events in your town since most of your time will go towards maintaining your lifestyle and tending to your homestead. Besides, the commute to and from your home will be too long and not worth it.

7. Uncontrollable Factors

We have discussed this previously. You must be prepared for the worst-case scenario, which is that everything that can and might go wrong. Sometimes something will happen you weren't mentally prepared for, and you won't know how to deal with it, despite the contingencies you have put in place.

8. Too Much of a Good Thing

Spending too much time all by yourself will give you a case of cabin fever, or at the very least, it will make you quite antisocial. To compensate for your lack of social life, you will throw yourself in your work, which will drain you, causing a burnout.

9. Money Problems

We're not just talking about the start-up costs. If you're not working a full-time job, things will get a bit tough for you when you must spend money on new items for your homestead. From where will you generate your income? Part-time work? Remote work? Have you considered this aspect thoroughly?

Chapter 4: What You Need to Get Started

As with everything, getting started is the hardest part of off-gridding. It's quite a drastic shift in your life, isn't it? The mere notion is enough to make you reconsider something as daunting as going off and living in a remote wilderness. The number of things you must do can also get quite overwhelming. Where do you begin? Do you start by learning woodworking and carpentry, or do you learn how to preserve your food?

To make this shift less scary, let's produce a checklist that shall modularize the whole thing into manageable steps.

The Initial Planning Phases

In this phase, you should, first, get all your financial affairs in order. Let's begin by making a budget that you'll spend on your new lifestyle, then moving on to reducing all unneeded expenses, such as buying new clothes, eating at fancy restaurants, and paying your Netflix monthly bill. Once you have done that, you'll notice that living off-grid doesn't seem as difficult anymore. Good. This is why this is the first step. Second, you should save money from this very moment. If you have any debts, pay them all off before going off-grid.

Once the budget part is out of the way, research on which off-grid style serves you best. There's a wide variety of styles, two of which we discussed in the first chapter, and the rest of which we will discuss later. It's all about finding the right balance between going off-grid and staying somewhat in touch with your old life. You're going to have to keep up with your relatives, after all. Sooner or later, you will also need tech to rely on. So, doing research about which style suits you best is important.

After that, start researching the local laws about off-grid living. Different states have different laws so consult with a lawyer beforehand, and if the laws don't allow you to live off-grid, it might be an excellent choice to move away to another state or even another country. You're going off-grid. It doesn't matter where you are, does it?

Now that you've researched the lifestyles and done your budgeting and decided which style suits you best, decide on how far you want to go off-grid. Do you want to go completely primitive, or do you want to stay in touch with technology? What elements are you willing to incorporate, and what elements are you willing to let go? Can you live without electricity, sewage and water? What are your workarounds for that? Think about it all before you take the plunge into your new lifestyle.

If you have health issues, consult with your doctor, and ask them how your health will factor into off-grid living and how frequently should you visit the hospital for regular checkups.

Ask yourself again, why are you going off-grid? Are your reasons strong enough to warrant a tectonic shift in your lifestyle? Are you motivated enough? What's holding you back from going off-grid? What are you going to do once you've successfully gone off-grid?

List all the skills you already have, such as woodworking, gardening, carpentry, and first-aid. Now that we have those out of the way, we can focus on the skills you don't have and need to learn to live off-grid, but that comes later. Remember the importance of baby steps.

Last, figure out a way to earn money once you've moved off-grid. Are you going to sell the produce from your garden? Do you intend to become a freelancer? Will you commute to and from work every day? It's completely up to you, but you must get it out of the way in the initial phase of your planning.

Learn How to Produce Own Your Food

First, gardening is an important thing you should learn about your off-grid lifestyle. Read about it online; follow YouTube channels that teach gardening and then start applying your knowledge in your current home. For starts, pick a few plants and plant them in the flowerbed and tend to them for a month to see if you can do it or not. It's not that hard of a skill – you will surely be able to do this! There are various gardening methods to choose from, such as greenhouses, aquaponics, hydroponics, and permaculture. Do your research about them and select one that suits you best while also being the most convenient for you.

Second on your list should be *hunting*. Are you interested in sourcing your food through hunting? If so, the first thing you need is a hunting license, the second thing is the gun itself, the third is practicing with the gun at a shooting gallery, the fourth is learning about animal tracking, and last, how to butcher animals.

If you don't want to hunt and instead want to raise animals for your food needs, visit a farm to see how they're doing it. If you are open to the idea, you can volunteer at that farm to practice raising animals.

Buying the Property

Pick out the area where you want to live, gauge it by checking its accessibility to water, the proximity of the neighbors, soil quality, and available natural resources on your property.

If you're satisfied with the land that you've chosen, go ahead, and meet a real estate agent and buy the land. You'll be surprised how relatively cheap it is. This is because, as compared to urban areas, the rural property doesn't cost as much, which is another reason the off-grid lifestyle is very appealing.

Planning Your Home

The first thing you want to do is consult with an architect or an engineer specializing in green homes or *eco-friendly* homes. It'll do you good to learn about green home building materials and methods. Once you've done that, it's time to decide the size of your home. If you're strapped for cash, consider building a house that is small in the beginning but one that can be added to later – perhaps in modules. Now, decide whether you want to hire people to build the house or if you want to do it yourself. Doing it yourself will teach you a lot of skills that will come in handy later - skills such as carpentry and construction.

Then, plan your property's layout in terms of exposure to the sunlight, water access, the location of water resources (well, lakes, rivers, streams), the gardens, waste disposal areas, and outhouses/outbuildings.

Plan the water supply system and plumbing by researching rainwater harvesting systems, hiring a geologist to locate wells on your property, and determining where you want your latrine/septic tank placed.

Next, we move on to the electric system. The first thing is a no-brainer. Do you still want to be connected to the grid or not? Of course not! Why else are you pursuing this lifestyle? Decide whether you want to use wind power, solar power or a combination of both. Do your research and consult engineers (solar power engineers and/or wind power engineers) about your equipment needs and your energy requirements.

Now, we move onto the heating system. What kind of insulation are you planning on using? What kind of heating system do you want? A wood furnace, a wood stove, solar heating, power generator, animal dung, or coal? Specify it in your checklist. To buy fuel, you should secure access to it by now.

You're also going to have to build a root cellar for the storage of food, particularly root crops.

What are your transportation methods? If you do not know the mechanics of automobiles, I suggest that you brush up on basic mechanic skills, since you will need them at your off-grid setup whenever a vehicle malfunctions. You won't always have a mechanic at hand, and you're striving for independence; it makes sense to learn auto mechanics. Now, consider what kind of vehicles you can use at your property according to the terrain. Examples include bikes, quad bikes, motorbikes, four-wheel drives, and trucks.

It's time to reevaluate your budget so that you can plan for permits, labor, building expenses, machinery, repairs, and other costs. Prioritize your costs. It's recommended that you produce a project plan for your home, itemizing each item and its expenses. Last, consider which projects are essential (they should be your priority) and which projects can be delayed.

If you will work from home, set up a framework for your off-grid office.

Moving into Your Home

The first step to moving into your new home is its construction. Decide when and how you will move in. If it's difficult to move in right away, you can move in at a later time and still give your time to your site by visiting on the weekends and setting up your camp or staying in an RV to ease yourself into this new lifestyle.

Plumbing and water are critical to off-gridding. First, install a toilet system, and then move on to drilling a well. You can set up a rainwater and greywater harvesting system as well as setting up a solar-powered shower to give you that warm bath at the end of a long day.

Install your heating system and stockpile for the winter. That includes firewood and fuel.

Now comes the electricity bit. Install a solar system as your primary source of power and buy a backup generator just in case you need it. Start practicing a reduced energy lifestyle so that when you move in, you're not overwhelmed by the scarcity of power.

Manage your waste by starting a compost pile and learning how to reduce waste and reuse it.

Starting your garden is another crucial step you need to take. Calculate how much food you will need for the winter and start planting accordingly. What kind of food do you want to grow? Make a list and then get started on designing your garden. Again, reevaluate your budget so you can afford tools and seeds. Make sure you plant your crops in the correct seasons. If you want nuts and fruits, you can plant an orchard for it. For your first crop, start by sprouting seeds. Learn how to repair and maintain your gardening tools.

Get chickens, build a coop for them, buy heat lamps, buy chicks, and raise them indoors. Then, move them to their coop when the weather is favorably warm.

Then, get livestock and build pastures, fences, and pens to keep them in. If you need a barn, build one. Visit livestock fairs to get an estimate on costs. How are you going to transport your livestock to your off-grid site? Also learn veterinary care for your animals.

Start Building Your Greenhouse and an Aquaponics System

Keep your health in check by learning first aid and natural remedies, and create/maintain a medicinal garden at your site. To avoid diseases, eat healthy food, stay active, and make regular appointments for annual check-ups, as needed.

Finally, be happy and content knowing that you have started living off-grid and have accomplished a goal that most people only dream of – but never actually pursue. You, on the other hand, have made your dream a reality. That's something to be proud of. You've earned bragging rights; go ahead and flaunt it.

We have only discussed the basics of what you need to get started. Now we will study each aspect in detail.

Chapter 5: Creating a Homestead Layout

After covering what we need to get started, let's get into a few layouts more extensively so you can decide on which homestead arrangement suits you best. We'll be covering around 28 of them to ensure we cover the topic exhaustively. Let's get right into it. For reference to the images and design, we are using this link:

https://morningchores.com/farm-layout/

1. The Tiny Backyard Layout

This is the most suitable option for those of you still living in the suburbs and yet want to be off the grid. In this blueprint, you will be harvesting your food in a back garden or on a small piece of land you own. You will set up a chicken coop in one corner of the place, vegetable boxes in the center, fruit trees by the eastern wall, and herbs by the western wall. A stone pathway can lead to your small backyard layout. The biggest advantage of this design is that you don't have to move to a new place. You can still be off the grid while living in the suburbs.

2. ¼ Acre Layout

This is for those of you who own land in a rural area. How much land? Well, a quarter of an acre. Suppose you're confused about how you will design a blueprint that suits your purpose. It will not only give you space to grow your food but will also have plenty of room for livestock such as chickens, sheep, rabbits, and goats. You can start by dividing the property into around five modules. The topmost area of the land should be a goat pasture for your livestock grazing; below it, add another goat pasture where you can plant fruit trees and a chicken tractor. To the upper left of it, create a goat pen, and below that, a chicken coop. To the utmost left, place vegetable boxes for herbs and plants. Does that sound manageable?

3. ½ to 1 Acre Layout

There's plenty you can accomplish in this amount of area. Not only can you raise your livestock, but you can also plant all the plants you could need, and a little play area for your children.

In just a half to a whole acre of land you can add nut trees, goat and sheep pastures, a chicken coop, fruit trees, pens for your goat and sheep, a compost pile, rabbit and chicken tractors, alfalfa, a tool shed, vegetable boxes, an herb garden, and your house. You'll still have a space left over that you can turn into a calm and serene orchard or garden.

4. 2 to 3 Acres Layout

If you have bought two to three acres of land for your off-grid homestead, it means that you're considering this seriously. Good for you. The biggest advantage of this much land is you can plant oats, wheat and alfalfa while also building pens for pigs, goats, and sheep. In two to three acres, you're also able to build a medium-sized barn for your animals. Besides the barn and the pens, you can add rotating pastures, a garden, a shed, a coop, and an orchard. You'll be able to build a good-sized house in that much land as per your requirements.

5. Urban Permaculture Layout

Permaculture is a design principle that centers on whole systems thinking, directing, and simulating the conditions and features that exist in natural ecosystems. At least that's what Wikipedia says about it. If you're not planning on moving out into the wilderness and intend to go off-grid at your present home, this design will be most beneficial for you, but there's a caveat at play here. You're going to have a lot of land at your disposal. This design is very organized, very structured, and well thought out. There's a décor element to it, as well as functional utilization. You can plant anything you want, from an orchard to a garden, in this design.

6. Edible Garden Layout

If you live in an area with a warmer climate, this design would be a clever idea. You can grow your pomegranate trees, citrus trees, vegetables, seeds, and nuts. The edible garden is a sustainable and functional design.

Here's how it works out: in the center, you have a house. Above it, there's a patio extension. Above that, there's a parking lot with ample space for two cars. To the right of the house, you can set up a garden with figs, avocados and oranges. To the left of the house, there's supposed to be a bed for herbs. Further left, you can grow lemons, oranges, and mandarins. There's a shed beside that, above which you can plant apricots, nectarines, hazels, peaches, and apples. This design is the best option for vegetarians. Speaking of which, put a greenhouse, a wood store, and a toolshed on your land.

7. Designer Micro Layout

This is yet again an awesome design that shall appeal to you if you're a vegetarian or vegan. This design is not only functional but also decorative, with a big house in the middle and plants and trees surrounding it from all sides. Loquat, hawthorns, and ivies can grow on the right side of the house. You can add a quaint deck to the far right, overlooking vegetable crops. The layout's top wall should contain all trees, ranging from pomegranate to pears. Cane berries, Nashi, quinces, apples, holly, apricot, and cherry trees can be planted on one side. To the left, you should have a pond around which you can grow thyme, dwarf lemon, and Mediterranean herbs. The only drawback of this design is that there's no place for livestock—only fruit, and vegetables. Although on the plus side, you have a very ornamental garden at your disposal with a ton of fruits and vegetables available in both on and off-seasons.

8. Designer Chicken Layout

The designer chicken design is optimal for raising chicken in a convenient and neat manner. This homestead style is aesthetic and functional at the same time. You can have a longhouse with a carport on one side, a patio on the other, and two doors—one for the kitchen and the other for the laundry—and all the plants and trees that we've mentioned in the previous sections. You can also add the main cropping area for your crops. If you have pets such as cats or dogs, you can add a fence around their area so they don't interact with your chickens. The chicken shed is on the corner of the homestead, with a butchering station next to it, though not so close that your chickens can see their fellows being butchered.

9. All-Inclusive Layout

Are you ready for the big one? Then look no further. In this design, you will have everything, and by everything, I mean *literally everything* your off-grid homestead requires, from sheds to nurseries, from plants to pens, from trees to tracks. *Everything.* Of course, it shall require just as much space, but in return, it will make you completely self-sufficient. Not only is there a placement detail for each and every single element of the homestead, but there's also an aesthetic element to all the placements. It also includes a design for your own power and water system.

10. Basic Farm Layout

The basic farm layout has everything that a farm constitutes, from barns to stables and silos. There can be cabins you can use as a granary and a woodshed. You can dig for a pond you can use as your water source in dire situations later. Essentially, this design is perfect for you to emulate a farming lifestyle. There's just the added requirement of a large amount of land for this form of design, as it can't fit in less than an acre, but you can take the layout and remove things from it to suit your purpose, thus accomplishing it in a smaller area.

11. The 1-Acre Dream Layout

It's a dream layout because it is completely independent of the amount of area you have. You can modularize it into smaller chunks and fit everything well within an acre. The homestead will be functional and compact, fulfilling your every need without making you exert yourself by traveling all over your property. It's manageable, dependable, and sustainable. You can plant vegetables in one area, crops in the other, build a coop for chickens, a shed and pens for livestock, an orchard for fruits, and a greenhouse. It depends on what you want instead of the other way around. You'll also have your choice of home styles to pick from, which we'll discuss at length in the next chapter.

12. Permaculture Layout

The previous permaculture layout we discussed was an urban one. This is different because it requires a lot of area for your permaculture. This is your best option to live in a manufactured home as opposed to, say, a cottage or a hut. You can plan to build this design all-around your house. This form of layout is modeled after an old-style farmhouse. It has versatile features and maintains an aesthetic sense exclusive to permaculture farming. You can also use your plants as natural hedging for your properties, which is an effortless way to set up boundaries.

13. 1/10th Acre Layout

It's a simple, elegant layout with seven main components in a grid-like setting. The first section will be on the top left for vegetables with eight beds each with each bed being 4 x 8 feet. The second section will have fruits and nuts in it, including multiple fruit trees, vines, berry canes, and strawberry beds, all of which will be along the fence line. Third, there shall be a bed for herbs beside your home. There isn't a lot of space for grains, so we're excluding them from the list. You can have a smalltime poultry pen with six chickens in it, and you can keep rabbits as well since there isn't room for larger livestock here. In terms of wild food, you can add a beehive or two for honey.

14. The Garden Layout

This layout is not exactly a homestead layout, but consider it if you want to add a nice garden to your existing house. That way, you will become self-sufficient in terms of food, if not other aspects such as electricity and water. The way to sufficiency is gardening and livestock. We'll discuss the power and water part in detail in the upcoming chapters.

Having a garden layout will provide you with all the food you could need, giving you your daily dose of minerals and vitamins, ensuring your healthiness. You can sell surplus produce at the local farmer's market to generate a passive source of income.

15. Complete Design Layout

This layout has everything an off-gridder could need: a nice three-storied home, a sign by your home to advertise that you sell produce, a storage shed for tools, a relaxation corner, a hundred feet farm, a compost area, laid hedges, deep beds for small space gardening, a mini-orchard for fruits, a sturdy fence around your garden, a house for ducks or rabbits, a pond for the ducks and fishes, a beehive for your family, and last, a grazing area for pigs and other livestock. They thought of everything when they made this design!

16. The 2-Acre Homestead Layout

The two-acre homestead contains everything that you need in as little land as possible. For most of you, two acres would be more than enough to sustain an off-grid lifestyle; while for others, it might be too small. No worries, we're discussing all the possibilities here. For reference, two acres is roughly 90,000 square feet. You can check out the detailed infographic on the website. You can adapt a lot of plans for your land and customize the individual elements to suit your needs. This design is very thorough in explaining where to place each element.

17. The Hamilton Permaculture Layout

This layout has something that we have not considered before: a worm farm. Worm farms will give you great compost, which in turn will be advantageous to your gardens. There are bees, chicken tractors, aquaculture, and guilds in this design as well. The barebones of the design start off with a house, behind which is a sustainable backyard, next to which there's an adobe, a stacking, a compost pile, and a bed for gardening. The design has a luxurious feel to it, ergo the presence of "Hamilton" in the name.

18. Larger Farm Layout

This is perfect for those of you who have more than a few acres of land on their hands because it not only includes all the common farm buildings, but it also has a functional silo too. In this design, there is an orchard with a beehive and fruit trees, a pigsty, a bunker for the silo, the silo itself, enclosures for animals, a greenhouse, vegetable gardens, permanent pasture, a fallow, a meadow, an area for crops, and ornamental trees.

19. Urban Homestead Layout

This layout is a clever idea for those of you considering an urban design. There are functional areas and rest and relaxation areas on this layout.

20. Old School Homestead Layout

The old school homestead is reminiscent of the Wild West, complete with a ranch and all, but you must have a few acres lying around to pursue this layout.

21. All-Inclusive Small Homestead Layout

Perfect for multiple meat sources, this design doesn't rely on acreage as much as it focuses on functionality. There's also room for cold frame greenhouses. This layout utilizes all the space in a compact manner.

22. Family Food Garden Layout

Ideal for your family, this food garden layout is efficient because it utilizes all the space for growing a ton of food. If you have more than ample space available, it's probably best to keep the animals away from the house to avoid the smell and the sounds.

23. Family of Four Mini-Farm Layout

Most suitable for you if you have a small family, this design caters to a familial lifestyle, considering every aspect of homesteading that your family could require.

24. Modern Homestead Layout

This layout has an amazing planting space where you can plant edibles, fruits, and vegetables. Another huge plus is the modernity of the design, which will appeal to the esthetician in you.

25. Real-Life Layout

The real-life layout helps you in living a practical life, not that all the others don't. It's just that, realistically, you will have an easier time adapting to this design than all the other ones. You can easily manage all other aspects of your life without worrying about making too drastic of a shift, as other designs require you to do.

Now that we have covered every (well, almost every) layout for your off-grid home, let's move on to choosing your style of living structure.

Chapter 6: Choosing Your Style of Living Structure

Whether you want to reduce your carbon footprint or are simply interested in moving off-grid to live your life in secluded solace, you will need a house plan. A house plan differs from the layout because it focuses on the interior of the house. There are a lot of considerations to be made when living in your off-grid house, which we will discuss now.

Eight Features of Off-Grid Homes

The first thing you must remember is that off-grid homes must be efficient in terms of energy and water. There are many ways to go about it and we will discuss it in detail. It's all about the design of your home. Consider these eight things when choosing your design and style of living structure.

1. Insulation

The off-grid home should have very thick insulation, or at the very least efficient, if not thick insulation. It should be better than an average traditional house's insulation so you can conserve the sun's heat in the winters and keep your house cool in the summers.

2. Eaves

Eaves help you with heating your house when it's cold and with keeping it cool when warm. Practically the same function as insulation, but with a different method. Overhanging your roof so the winter sun—at its low angle—can get through your windows while the high angle sun's rays will not only reduce the energy consumption in terms of cooling and heating but also ensures that you get plenty of sun for your Vitamin D. The eaves shall be applied to the house's south side for those of you who happen to be living in the Northern Hemisphere.

3. Solar Light Tubes

Again, this is one of the best ways to save costs while remaining efficient. You can harness the power of the sun during the daytime to light up your home. Sure, the solar tube system will ensure the lighting for the daytime, but they're not intended for nighttime lighting. With that being said, it doesn't mean that they're not advantageous to have; they are, after all, saving you a ton of money in terms of power consumption, even if it is just for the daytime.

4. Cisterns

You need something to collect the water with, so why not cisterns? This way, you'll have water on your hands whenever it rains. We're not suggesting that you use that for drinking; but should you decide on using it as drinking water, filter and boil it first to purify it. Place plenty of cisterns all around the house to increase the surface area for rainwater storage.

5. Water Heater

You can either choose solar water heaters or tankless hot water heaters. The tankless ones only turn on when you need them to and promptly turn off when you're done consuming water for bathing or cooking. To step it up a notch further, you can choose solar water heaters for heating your water in the daytime. The only downside to this is that you won't get to use hot water at night, but that's okay, isn't it? You can take your baths in the morning ... no big deal.

6. Stove

A wood stove is a must-have in the arsenal of an off-gridder. It shall serve as the centerpiece of your house. Wood is probably one of the most effective ways to heat up your house, especially in those long, chilly winter months. When choosing a stove or, even better, building one, you must make sure that it's made of stone and is properly insulated to give you the most efficiency for burning wood. This is especially beneficial when, such as in winters, you have a short or limited supply of wood.

7. Ventilator

Not the kind that hospitals have. This ventilator is supposed to function as a heat recovery mechanism by capturing residual heat before it gets a chance to leave your home.

8. Solar Panel

If you install solar panels on the south-facing side of the roof, you'll get yourself a ton of solar power at your disposal. Remember, we're aiming to maximize our renewable power sources as opposed to other forms of sources, such as diesel. By placing the panels on the south side (no, not Chicago), you can get the most out of the sun in all kinds of weather.

Off-Grid House Plans

Whether you're looking to get started, or have already started with a modest budget on your hands, these following plans will give you a direction and an idea of how much budget you need. Let's get right into it.

1. Prefabricated ARK Shelter

The ARK shelter is perfect for those of you on a budget, as it comes at a relatively low price of $50,000, which includes the transportation and installation. Let's be honest with each other for a second. As an aspiring off-gridder, you probably have no idea on how to get started building your own home. That's okay. We don't need to reinvent the wheel when we have prefabricated options at hand. There are multiple styles that ARK offers. The most viable one being a cozy shelter designed for two people along with a cabin that's the length of a shipping container. This particular design also has a living room in the center, with just about enough space to fit a small dining table in there. Other key features of this home include a kitchen, a bathroom, and a bedroom. There's a huge panel window on one side, from which you can admire the view of nature.

Since it's a prefabricated setup, it comes with its own installed solar panels and can also utilize wind power should it be needed. There are cisterns on the roof to help you collect rainwater. There are different options for your sanitation, including septic tanks, waste chemical treating, and so on.

2. Minim House

The tiny house movement inspired the Minim House. You can use it on the grid as well to wade the waters before diving in. The house is small, with only 260 square feet, but there's room enough for practically everything. There aren't any walls inside the house, so the whole place gives off a studio apartment vibe. The bed can be pulled out of the wall and pushed back into it, thus giving you ample space to move things about. The kitchen layout of the house is roomy as compared to the rest of the house.

A series of solar panels on the roof provide a power of over 900 watts in terms of collecting energy. There is also a set of batteries built-in the structure, giving you the freedom to be totally off-grid. The batteries alone aren't enough, so you will need additional power sources including solar power, as discussed, and generators.

Unlike the ARK home, there are no cisterns in the Minim house, so there's no way to collect rainwater, but there's a 40-gallon tank that's tucked away somewhere below the couch. There is also a water filtration system attached to that tank. There are big glass windows that will be the primary source of heat in the daytime. The doors serve the same purpose. The one downside—other than the lack of cisterns—is there is no wood stove to heat the house centrally.

3. Off-Grid Shelters

Suppose you are still a little apprehensive about the process of building your own homestead while at the same time not in favor of pre-fabricated homes, no need to worry; there is a third option for you in the form of off-grid shelters. It's recommended that you consult with an architect or a building firm that specializes in this exceedingly kind of shelter. You have a book of plans at your disposal, which the firm shall provide. This will inspire you to be creative with the interior design of your shelter and not just the interior décor. You can also be quite involved in the exterior design, tailoring each part according to your aesthetic and choice.

4. Tiny-A-Frame

This is your chance to have a trial run at the off-grid lifestyle without investing too much of your money, as the Tiny-A-Frame costs less than $1000, particularly when you're using recycled material. The cabin, as the name infers, is tiny—uncomfortably so— and covers just about 80 square feet. There's only enough space for a bed in there. There aren't any kitchen or bathroom units, so you must depend upon exterior resources for both. There are solar panels at the top of the frame house that provide little energy, but seeing as how this is such a tiny home, don't be worried about that. Like the previous house, the heat is provided by the windows and the doors. You won't be able to occupy this space in the winters, as the lack of a central heating system will make it difficult to bear the harsh weather.

5. Penobscot Cabin

This is the best plan for you if you are feeling confident about building your own house. Not only is this plan easy to build, but there is also so much potential in the way of adding new modules and areas to it once you have started off your homestead and have been running it for a few months. By default, this cabin does not come with cisterns or solar panels, but you can add them later. Other than that, there are no large windows either. Since there's going to be wood used in constructing this cabin, you will have natural insulation, which will come in handy in the winters. If you're skilled in terms of architecture and building, you can add on more modules once you've become comfortable building upon your cabin.

What About Underground Houses?

If you admired the books and movies featuring Hobbits, you will absolutely love the idea of having your house in a hole in the ground or under the hill. If you think that the homes would be dingy, dark, and smelly, you're mistaken. There are as many types of earth-sheltered homes as there are of the aboveground homesteads. Let's look at three of them for starters.

1. Earth-Covered

These types of houses have a living roof above them. They're going to look very much like normal homes, but without any roof. If you've got an earth-covered home under the hill, it shall serve as a cave of sorts, covered from three sides by the hill, and the fourth side shall have a door and windows on it.

While it is not entirely underground, this home is a primer in getting you started in living the underground lifestyle. It will ease you in, so to speak.

2. Earth-Berm

The earth-berm houses enjoy the popularity factor. Their styles vary from house to house. Some houses might have half-covered walls, while others might have two to three sides covered by the earth. Mostly, these houses are set in hillsides like Hobbit-holes and look quite regular when seen from the front. If you're the kind of person who doesn't like the aesthetic of subterranean houses, this might be your best bet.

3. Subterranean

This one is completely underground and has none of its walls made of brick and mortar. All four sides and the roof are covered by earth. Most of these houses are built into hills, just like earth-berm ones. They're an awesome place to stay, but again, one style might not suit all. Note that most of these subterranean houses offer you the luxury of traditional homes within the ground. From the inside, minus the windows, you will notice nothing that will suggest that you're living underground. Another advantage is natural insulation. You can save on heating costs by adding deep insulation to it. There's also the option of adding windows on the side with the door on it.

Advantages of Earth-Integrated Houses

1. Save on Power Bills

Depending on where you live and the weather there, you know how exorbitant air conditioning and heating bills can get. When you have an earth-sheltered or earth-integrated home, you will end up saving a fortune on your bills, since it will be buffered from all sorts of harsh weather from all sides. There's going to be coolness in the environment in the summers and hotness to it in the winters. It's a win-win as they say, right?

2. Natural Protection

Having a subterranean house will give you natural protection against most calamities, such as tornadoes and hurricanes. Since you're practically underground, there's little chance you will be affected by those calamities. That alone is the best and biggest factor you should consider when opting for an underground house.

3. Little to no Insurance Required

You'll find that insurance companies won't be taking a ton of money from you, as they know that there's going to be a lot of costs saved up in the form of natural protection. You're basically making their job easy for them. Traditional homes have a dozen natural threats posed to them, which subterranean houses do not.

4. The Cheaper Option

These houses don't need a lot in the way of architectural design and up-front costs, making them way cheaper than any aboveground homes. Also, since you're not using a ton of resources in the building process of the house, you're saving money on that front.

5. Cheaper Maintenance

Since this house has no exposed walls, there's no maintenance required. No repainting, no gutter cleaning, no refinishing, and no major renovations. They're not only easy to build, but they're also dirt cheap (see what I did there?) to maintain. You can save that money for designing the interior of the house.

6. They're Quite Unique

How many people do you know with underground houses, fictional characters excluded? You can spark conversations by telling people that you don't live in a traditional home and how it's better than their homes. There are online communities where people come together and contribute to the knowledge-bank of underground living tips and resources.

There are two things you must remember when choosing an underground house. First, there's going to be an extensive amount of effort that you will put in the construction and finalization of the house. Second, a lot of contractors and builders do not have experience in building these homes.

Chapter 7: Water and Sewage Options

Water and sewer are essential to a comfortable home, whether you live off or on-grid. There are more water options than there are for sewage, but all will be discussed.

Water Options

Here's a mindboggling statistic: the average person consumes anywhere between 80-100 gallons of water per day. Besides electricity and food, your main concern should be your water consumption. Realize that as someone approaching an off-grid lifestyle, you must remember the scarcity and the renewability of your resources. You're not going to have as much of anything as you did back when you were living in the city. The learning curve is a little too steep with this nugget of wisdom. Another thing to keep into consideration is an emergency. What will you do if an emergency occurs?

Let's discuss six options for your water supply and storage, so we're prepped for anything that nature throws our way.

1. City Water

Most homes built within the city will need to be connected to the city water system. You will have almost no choice in paying for the connection due to limited acreage and city requirements. Many HOAs will decide whether you must be connected to the city if you are within city limits. Mortgage companies are also particular on this point. Mortgage lenders want predictability when it comes to water. Unfortunately, it can come with downsides, such as chemicals to purify the water, being expensive, and becoming contaminated on a large scale. As someone who wishes to live off-grid, chances are you will not choose a homestead within city limits.

2. Wells

Wells are expensive to dig. It takes somewhere around $10,000 to dig one well, but they're self-reliant and independent of all sorts of sources. You're getting your water from the earth. Not the city. Not through pipes. Not through anywhere else. Plain, pure, raw earth water will always at your disposal. It's an excellent option to consider if you have a lot of land and have consulted with a geologist who, in turn, has told you that there's a water source running under your ground. A huge plus of this source is that there's no need for plumbing, no need for cisterns, or storage. All the water you need can come from underground, you can use it as much as you like, and the rest will remain reserved on the wall. It's an almost endless supply for a small cost. Did you know that your city water might be contaminated? This is hypothetical, but if it was to be, you're getting free from that contaminated water by choosing the well—which is as pure as water can get.

3. A Hand Pump

A hand pump is handy to have at your homestead for those times when you can't rely on your well all the time - or if it is too much work. Most wells come with pumps of their own, so there's no need to worry about having to install one all by itself. There are electrical pumps available with electrical wells these days, which are recommended. But if you don't have those available at your property, choose hand pumps, as there's no satisfaction greater on your homestead than doing something with your own hands. Invest in one close by your homestead, so you don't have to go all the way to the well to get water. It's one of the most budget-friendly options available to you.

4. Store Your Own Water

This sounds quite simple and straightforward, doesn't it? The truth is a commonly overlooked solution, so we must bring it up as the sixth point. This is critically important if you have a water source on your property, such as a river, stream, or a freshwater lake (even wells). You need to store all that water somewhere, don't you? We're not talking cisterns right now. We're talking about water jugs, containers, and other forms of storage that you can keep at your home. This isn't a long-term solution, mind you. This is a just in case for when there's an emergency, and you're cut off from all sources of water and power. Most of the off-grid living entails you preparing contingencies for when things go wrong. It's not necessary that they will, but just in case they do, you're prepared.

5. Rainwater

If you are living in a place with a lot of rain, this should be your go-to option. Now we're talking cistern and storage tanks, purifiers, filtration systems, and all that jazz. If you're getting constant rain, then this can be your long-term solution for water, but if you only get rain in spells, consider other alternatives. Be warned, though, that if you use that rainwater for drinking and cooking purposes, you will have to make sure that it's safe first. You must install a purification system specifically for this purpose. Always use this form of the water source with an alternative. This can't be the only source you rely on.

6. Hauling Water

Consider the worst-case scenario (yet again): your property is on dry, parched land with no well, no rain, and no other source of water. What are you going to do? Do you remember the bit where we talked about preparing for the worst? This is the worst. What are your next steps? Here's how you get started. Buy a huge tanker and attach it to the back of your car. You're going to take a trip to town and fill your tanker at the water station. You're going to take it and come back to your property and fill it in reserve dug in the ground. This is to keep the water protected from sunlight and to keep it cool. This is a tedious manner of work, but it's worth the work if you're in a land with scarce water.

Last, as a bonus, we'll discuss condensers. Condensers make water appear out of thin air. Literally, if you're living in an area that gets a ton of humidity, get a condenser to turn all that humidity to water.

Sewage Options

Living off-grid requires distancing from city water and sewer. If you are hooked into city water, then you will be sending your sewage to a city treatment plant. It is not all bad to use city sewer. Once you flush, you don't have to think about it again. Going with off-grid options requires more cost upfront, but you save on monthly payments.

1. Septic

A septic system either requires a vault or a leech bed or can involve both. Septic tanks or vaults must be large enough to sustain your household. For five people, a three-vault system is recommended. The type of soil you have also dictated if you need a pump that will get the sewage into your vault. Septic systems can run you as much as $10,000 but are often closer to $5,000. The price is always based on vault size, septic pump, and in-ground plumbing, plus labor to dig a hole for the system.

Once you have septic established, you have extraordinarily little maintenance to worry about. One of the oldest tricks in the book is to dump yeast in a toilet once a month, and it will keep your vault or tank exceptionally clean. If you do not use something like RidX or yeast, you will need to have your vault emptied every few years. This can be costly.

2. Latrine

Whatever you want to call it—latrine, outhouse, poop house—it is a hole in the ground you dig and place a shack over to keep your butt warm and off the ground. Eventually, you need the outhouse pumped or to fill it in and move your hut a little distance. Going with this option means outdoor excursions—not ideal—in cooler climates.

There is nothing wrong with an outhouse if you don't mind going outside and getting it pumped out every so often. If there are no HOA or city limits laws to interfere, you can live off the grid this way.

3. Camp Toilet

RV toilets are another option. Unfortunately, this method requires you to empty the small holding tank quite frequently. You also must live near a campground or have a larger hole where you dump your wastewater.

4. Compositing Toilets

There are at least three composting toilets to choose from such as: Humanure, Clivus, and Commercial. The Humanure bucket is something you can create yourself or buy pre-made. The Humanure composting option has two stages, where you collect your waste in a pot or vault, like a latrine, but usually by using a bucket. A standard bucket with a toilet seat and wood box to hide it plus an enclosure with a small fan for a vent ensures you have comfort without the smell. You will need to cover material to help keep the smell down. Woodchips, sawdust, peat moss, dirt, or clay can absorb the liquids and reduce the smell. About two times a week, you will need to add the bucket contents to an outdoor compost pile. After you dump the material, it will need time to cure, and then you can use it for your garden. This option can cost you as little as $20 for the bucket.

Clivus composting toilets range in cost from $100 to $1,000. They are pre-made compositing toilets made for those who wish to live off-grid. It works in the same way as your homemade option, where you have a waste container with an absorption material; but you can build it into your home so you can empty the toilet less often by placing a wheelbarrow underneath.

Commercial composting toilets are more than $1,000. They are plastic and like the Clivus. You can get about 80 uses out of the toilet, dumping the liquid once a week. Typically, peat moss is the best system, with a urine separator attached. You need less peat moss because you separate the urine. A few commercial options include Nature's Head, Separett, and Sun-Mar.

5. Incinerating Toilet

If the idea of dealing with your waste is not favorable, consider an incinerating toilet. You do not need an expensive septic system or a way to empty your solid and liquid waste each week. An incinerating toilet, like EcoJohn and Incinolet, will turn the waste into ash. The downside is the energy it takes to incinerate your waste. There are gas and electric models. Both types will burn the waste until there is just a little ash left, which you need to bury. The burn cycle takes about an hour, but you can still use the toilet while it is burning waste. If you are using solar power, you will need to account for higher energy usage with an incinerating toilet. For gas models, you will need more propane than if you went with septic, composting, or outhouse sewage options.

Of the above options, a septic system is the most effective for living comfortably off the grid. Animals go wherever they wish in nature. They don't even bury their waste. I don't recommend this idea because floods or other issues can create problems. You are better off with some type of vault system, even if you compost your waste.

Chapter 8: Heat and Electricity Options

Planning for heating and electricity is a smart choice regardless of where you will build your homestead. You want to decide on the heating and electricity for your home to ensure you are designing it for optimal energy efficiency and comfort. You will have plenty of choices for heating, although some are more cost-effective than others. We will begin with heating options before talking about the two most common off-grid energy options.

Heating Sources

Commonly, electricity, gas, oil, and wood are used to heat homes; however, you do not have to stick with those choices. Instead, you can choose among numerous options to heat your environment and ensure your electricity, gas, or oil usage is kept at a minimum. Living off-grid is not only about creating a minimal impact situation but also about low-cost.

1. Electric Baseboard Heaters

Electric baseboard heaters rely on a source of energy, wind-powered or sun derived. As an independent system for each room, you will need to determine how many heaters per square foot are necessary to keep the room warm. Account for any sunlight that can improve the warmth of the house. Small rooms like bathrooms may need one heater, while larger bedrooms and living rooms require two or three heaters for an even distribution of warmth.

2. Central Heating and Air

A heating and air system, usually called HVAC, has a central furnace, which pushes warm or chilly air throughout. The air goes through metal vent shafts and is distributed into each room. It has one thermostat, which can have an energy-saving property, where you set the temperature for when you are there or away. These systems are costly and require more energy to run than some of the other options discussed in this section. You will need to provide yearly maintenance and change out the filter every three months.

3. Boiler

A boiler works with propane and water. A boiler works like an electric furnace pushing air throughout the house. The water will evaporate as it is heated to create warm air that will then go through pipes or metal vents to heat each room in a house. Boilers require yearly maintenance to ensure carbon monoxide poisoning will not occur. Like central heating and air systems, boilers are expensive to install.

4. Corn or Pellet Stoves

Corn stoves came to be during the depression when coal was too expensive to buy. Families burned corn and found they could heat their living space for a lower cost. Corn burning or multi-fuel pellet stoves provide 8,000 BTUs, which is like the wood pellet option available. Depending on where you live, you can find corn is more cost-effective than wood pellets.

5. Masonry Heaters

Masonry heaters are like wood stoves; however, instead of needing a fireplace, you build a brick stove, with chimney, and burn wood directly on the stone. Wood stoves can lead to a creosote fire, whereas masonry heaters reduce this risk. Masonry heaters use a renewable fuel, they can fit your home size to ensure you get enough heat, and they are easy to operate compared to a wood stove. Masonry heaters are also exempt from fire bans because they burn cleaner. There is little maintenance required, only the occasionally emptying of the ash and cleaning the chimney. The downside is that you need to be there to monitor the fire, so it is not usable while you are away. You also need to make sure the build is airtight to get the most out of the heat produced.

6. Fireplaces

Fireplaces or wood stoves use wood to create heat. You usually need to install a blower, so the heat radiates further into a room. They work great in the room they are in or if you have a two-story homestead with an open floorplan upstairs. You need to be home to be safe, and you are subject to fire bans. You will need to clean the fireplace of ash frequently based on how much you use the stove. The chimney will also need to be cleaned regularly.

7. Air to Water Radiant Heat

Radiant heat is not a new concept. Like the boiler, you use a system to warm the water. The difference is there is a radiator, typically, in each room that uses water to produce warm air.

8. Solar Window Boxes

Solar window boxes warm the air instead of vents, which is then pushed through the house like central heating and air system. The key to solar window boxes is to make sure the boxes have direct sunlight. You will need a translucent box that will warm in the sun. The warm air will enter the home via the window or other opening and heat a room. You would need one in each room. You should not depend on this as your only source of heat, depending on where you live.

9. Home Design for Heat

It might seem counterintuitive to have more windows in your design. You might think the frigid air will make it too cool and take away from the heat you want to keep inside. Depending on where the sun hits your home, having a wall of windows can ensure you have more heat coming in during the day. Like the window boxes for solar-generated heat, a wall of windows in direct sunlight will heat that room or multiple rooms on that side of the house. Rather than running heaters and using more energy, you can use the natural warmth of the sun.

10. Biomass Systems

Using your composting pile, you can also use the sun to generate heat. A composting heap in direct sunlight will become very warm. If you have a biodegradation setup with coils of pipe, you can feed the warm air generated by the compositing pile directly into your home. The biodegradation system ensures pure air reaches your home. The amount of compost you have or the number of piles you have determines how many rooms you can warm-up.

11. Using Natural Landscape

Some of the most interesting tiny homes, with off-grid setups, use the natural landscape around the house to generate warmth. For example, if you build into the side of a hill or mountain, where dirt and rocks basically cover one portion of your house, you can keep more heat inside. The design requires concrete or masonry work. One house even created a roof garden, allowing food and even moss to grow. The sun heats the material and, in turn, keeps the heat inside. With windows facing the sun, the home with its concrete walls and the covered roof kept the owners warm. You can also do this, depending on where you want to live. Concrete, adobe, or brick homes bake in the sun, which helps keep the house warmer on the inside. Darker stain or painted walls also help generate more heat in the sun, which can ensure the inside is warmer when the sun is up.

12. Oil Based Heaters

Radiators, oil lamps, and oil heaters are three ways you can generate heat by using oil instead of propane, wood, or the sun. To heat your home while you are away, you need an oil radiator. Oil lamps and heaters should be used only if you are home unless they are similar to baseboard heaters.

Electricity Options

There are two ways you can live off-grid and produce your own electricity. It does not mean you cannot live off-grid with just these two options for light and heat. Oil lamps are one way to have lights without electricity. Candles are another. The drawback is the harder it makes a living. When cooking without electricity, it means lighting a fire in a woodburning stove or making a campfire. You could have propane for cooking, and that would also allow you to use propane methods for lights and heating. However, let's assume you would rather use one of three energy power options.

1. Solar Power

Solar power is not as simple as installing panels on your roof. Solar power involves photovoltaic panels, an inverter, and batteries. The batteries store the energy you create with the photovoltaic solar panels. You also need plenty of sun or indirect sunlight. If you live in a place that does not generate enough sun energy, you may find you do not have enough electricity in your home to warm it, cook, and use lights.

While most of us want to be off-grid, and without a mortgage, you may decide a mortgage is necessary to get the funds to build your home. You cannot be totally off-grid with a mortgage. Banks are unwilling to lend to someone who uses solar power as the only electricity source. Keep that drawback in mind as you plan your home.

Before you can get a price for your solar power system, you need to know what components will use the energy source and what they draw. In electrician terms, you need to understand the load and run time of appliances, water heaters, lights, and other electrical components in your home.

Once you have the components you want to run and the run time, you can calculate the watt-hour. You will add all the watt-hours for every electrical device together. Most systems will lose about 30% in the generation and storage of energy. So, with the total watt-hour, plus the estimated loss of 30%, you can then calculate the load.

There are different solar panels, including mono and polycrystalline. Check the size, quality, and type to determine which option best fits in your budget and maintenance needs. Your battery selection is costly, and you want to ensure you have enough battery to store the energy, while also choosing quality. The batteries can run $400 per battery, so you want something that will last 8 to 10 years, rather than shelling out $400 every 2 to 4 years.

The next components are an inverter and charge controller. The charge controller sets the voltage to ensure no one appliance or circuit is becoming overloaded. The inverter will convert the solar energy collection into usable electricity. You have a direct current from the solar panel that needs to become an alternating current to run your appliances.

A lot goes into solar power, but if you have enough panels and calculate your usage correctly, you can have enough electricity to power your home and even have some leftover.

It is best to have a certified solar power expert help you with the calculations and equipment choices or to purchase a guide that goes into more depth about installing solar power.

2. Wind Power

Wind-powered homes are secondary options that work great in places with less sun and more wind. Some cities do not allow wind power. HOAs can also have issues against wind power due to the less than favorable appearance of the turbines. If the wind does not blow, you will not have electricity. You also have moving parts on a wind turbine, which requires more maintenance and costs than solar power, but the benefits can outweigh other options if you have plenty of wind.

Like solar, you need to know the load to correctly calculate the size of the turbine you need to power your home. Size will matter. A 400-watt wind turbine will handle a few appliances, but not an entire home. 900 to 10,000-watt turbines on a 100-foot tower can handle an entire house. You could have two or more turbines, but most often, going with one that does the entire house is more cost effective than having multiple turbines and towers.

3. Micro-Hydro Electricity

Micro-hydroelectricity is not as often discussed, but it is still a choice that works if you live near plenty of running water. Micro-hydro energy derives from a stream, river, or another consistent running water source. Energy comes from the water that flows from an elevated level to a lower level into a turbine. If one has a good water source, the energy can last for 24 hours, 7 days a week, and ensure you are never without power. You also require fewer batteries as part of the system because of the constant production of energy. Unfortunately, if you do not have a stream or river in your backyard, then you cannot use this option.

Each of the electricity options requires you to have a home plan designed for the conservation of energy. You want to make the most of your home plan, so you get direct sunlight and warmth from the sun. You also need to choose appliances with a high rating of energy efficiency to reduce the energy usage and therefore load on your energy system.

Chapter 9: Gardening for a Food Supply

Gardening for your food supply is a part of living off-grid. There are several benefits to growing your own vegetables and fruits. You have control over how they are grown, what natural pesticides you use, and the ability to keep your food GMO free. It is important to start any garden with non-GMO heirloom seeds. They do cost a little more, but the benefit is worthwhile. As we assess this topic, we will examine how to create a compost pile, keep your garden organic, and control pests. We will also discuss the best types of gardens for drainage and maintenance, along with some foods you might wish to grow.

Optimizing Your Garden Space

The most important part of your garden will be the planning stage. You not only need to determine the space you have for gardening, but you must also assess the sun, soil, and size. The more space you can allot to a garden, the more you can grow, but you can also find yourself where you are overusing the soil.

There are plants that cannot be near each other. They will compete and die instead of growing into luxurious fruits and vegetables. Tomatoes, squash, beans, and cucumbers are at least four plants that grow upwards and will need stakes or trellises to ensure they will grow healthy.

Mint overtakes a garden. As an herb, mint will grow quickly and vine along with the earth, overtaking any space you provide, so it requires constant care to ensure it does not cover other plants.

Strawberries also vine and can continue to grow throughout a row, expanding until you have too many. So, the following assessment will be necessary to successfully grow your plants:

1. Look at your land.

2. Where does the area get the most sun?

3. How much square footage can you devote to the garden without losing space for your home or other things you'd like?

4. Do you want to build a greenhouse?

The above four points help you optimize based on space and weather. If you live in a cold, snowy climate, an outdoor garden may produce for three months out of the year. This may not be enough to feed your family for an entire year. A greenhouse is a clever idea for small gardens that need to produce year-round because you have more climate control inside.

Another option is to have a winter/summer grow table, where you will cover your garden with plastic sheeting in winter and shade cloth in summer.

1. Garden Size

The size of your garden depends on several things, the first being the amount of land you possess. With just one acre of land for a garden, you can grow enough food for 30 people for an entire year.

You may not need a full acre if you are just trying to feed a family of four and have food leftover for sales. You can monetize your garden, which is discussed in a later chapter. For now, consider the acreage of your lot and how much food you need to grow to feed just your family.

The optimal garden size allows you to grow what you need without creating a feeling of scarcity. For example, an entire 30-foot row of lettuce might be too much for one family. Six plants can be enough for a weekly salad for five people.

2. Types of Plants

Grow what you know you and your family will eat. For example, if you do not cook with parsley, don't grow it. If you are carrot eaters, make sure you have enough plants to account for how many pounds of carrots you eat per year.

The types of plants you choose should be based on garden size, growing seasons, and your preferences. In mountain areas, tomatoes, blueberries, and beans grow well, while warmer climates support more diversity such as raspberries, strawberries, lettuce, cucumbers, and much more.

If you will plant directly in the ground, choose hardy plants that can withstand the frost. Also, research what each vegetable or fruit requires for sun, water, and plant food. Some plants need more or less sun than others. You want to make sure you keep like plants next to each other without creating a situation of competition.

Chives, dill, tomatoes, and peppers grow tall, while mint and basil are medium height. Lettuce is a shorter plant, as is parsley.

You do not want to grow a tall or medium plant near a shorter grower as it can shade the other plants too much and prevent growth.

The growth time should also determine how you plot out your garden based on types of plants. Carrots take several weeks to grow from seed, and even after 100 days, you may not have a very large carrot bunch. Whereas mint grows within a few weeks and will continue to expand unless you trim it every week.

Regarding types of plants, it may be in your best interest to get a book or guide that discusses each vegetable and fruit best suited for the growing conditions before you put it in your garden.

Types of Garden Beds

Raise beds, inground, and container gardens are three types of gardens you can create. Choosing the correct style will ensure that you have enough plants healthily growing to feed your family. There are seven types of growing beds, plus container gardens to discuss.

1. Raised Beds

A raised bed garden is where you build a box, place your soil inside, and then put your seeds in the planter. It can also be a mound of soil that is higher than the dirt around it. The idea of a raised bed garden is to ensure proper irrigation and soil medium for the best growth potential. Usually, you have one raised bed per plant type.

A raised garden bed allows a deep growing area where the plants roots grow down and out. Another advantage has the beds at eye level, which helps you take care of the plants without hurting your back. A higher bed can also help you see the pests better. Native soil can be contaminated or improper for the types of food you want to grow. Raised beds to ensure you have the correct medium.

Raised beds are a permanent garden, where the soil is exposed to heat and cold. If you build your beds you'll want to have thin walls as well as an irrigation system to ensure proper moisture because raised beds can become dry more quickly. You also need pathways between the beds to help your garden.

When choosing raised beds, you want to decide the ideal height based on the types of plants you are growing. A height of 12 to 18 inches is typically ideal for things like carrots that need at least a foot of underground growth. You also need to determine the width and length based on the number of plants you need to grow in the bed. Rectangles are the typical design; however, you can make T's, ovals, or circles depending on your landscape and other garden beds.

Some plants need 12 square inches per plant, which limits how many you can put in a raised bed rather than a mound raised bed style.

The wood you introduce into your garden for raised container beds should be untreated. Treated wood can have chemicals that will affect your plants. You want healthy soil, so going with natural untreated wood you stain on the outside is better. A raw treatment like linseed oil is even better than the stain. You want something on the outside that will help prevent wear from the weather. Overtime, sun on untreated lumber can cause it to become compromised. Typically, if you just get sun only on the outside edge, it should not interfere with plant growth. There are some studies that indicate paint will trap moisture in the wood that will make it rot overtime.

People have used a variety of products like block, composite wood, and railroad ties to make raised container beds. You are better off avoiding all these, including galvanized metal. They all have products that could create chemical toxicity even though current research is not detailed enough to show such a thing has occurred. Tires are also something to avoid despite how handy they might seem. Rubber is not an excellent product to contain anything you will eat.

Many people also line the raised container beds with plastic, which can cause health issues. You need to purchase food-grade polyethylene plastic designed for gardens to ensure there are no issues with plastic toxicity. Plastic can be helpful for a barrier between the bed and soil; however, drainage must be a consideration.

Mound raised beds to eliminate wood issues; however, most people believe in using a weed barrier to keep weeds from interfering with food growth. You do need to get a food-grade barrier.

2. Clipping Beds

Within the raised bed category, you have clipping and plucking beds, which house two types of plants. You might integrate some of your plants, so you would want a clipping bed filled with plants you will need to clip, such as chives and mustard greens.

3. Plucking Beds

Plucking beds are those where you remove the top leaves or flowers from the edible plant to keep it contained, but also to take what you need to use for recipes or to eat. Plucking beds usually have faster-growing plants like broccoli, kale, zucchini, cucumbers, and fruit plants.

4. Narrow Beds

Narrow beds are better for carrots, beans, tomatoes, peas, radishes, and other root vegetables. On either side of the mound, you bury the plant to give them 12 inches space. The root grows the food you will eat in the soil so the plant will have short tops.

5. Broad Beds

Broad beds are better for larger vegetable and fruit plants like pumpkin, sweet corn, cabbage, and cauliflower.

6. Herb Spiral

An herb spiral is a vertical style bed, where you use both the grown and vertical space to create an herb bed. The circular creation ensures plants have more sun exposure than others. You will also get different wind and temperature exposure for the plants. You want to have the correct soil for the plants you grow, and due to the spiral, there is water drainage from top to bottom. So, plants that require less water should be on the top.

7. Vertical Planting

Vertical planting beyond an herb bed can include using trellises, pergolas, and fences. Some people even use the side of buildings to grow their species of plants. Peas, beans, cucumbers, grapes, and kiwifruit are commonly grown in vertical garden beds. You start with a mount and then place fencing or a trellis around the plant so it can vine up rather than out. You get more plants with vertical growth because you can fit a plant every 12 inches rather than worrying about vining plants that need ground space.

Container Gardening

Raised beds and mound gardening that go directly in the ground are the best options, but talking about container gardening is important for those with limited space or seasonal needs. For example, if your landscape does not lend to gardening because you live in the mountains with too much rock and sloping ground, containers might be the better choice. In colder climates, you might want the option of moving the containers indoors during the winter season.

Like raised beds you construct, you want to be careful of what you choose for your containers. Buy containers like ceramic pots from the gardening section at your local store.

Also, realize you are limited in the space you have for containers. You might get one plant, such as one blueberry bush in a 100-gallon container. Herbs can handle smaller pots. You can have six plants in ten inches of planter without having problems.

The depth of the pots used also must account for the root system. For carrots, a deeper container is necessary. Strawberries are fine in a long shallow container.

Hydro-Gardening

The most famous hydro garden is the Aero Garden. It is a fully contained system that uses water, plant food, and ten hours of light to help you grow tomatoes, peppers, lettuce, and herbs. They also have a strawberry Aero Garden. While expensive to get the container, the plants are affordable, organic, and the yield is high. But it is unnecessary to purchase the small gardens to have a hydro or water-based garden.

You can create your own hydro garden using large containers filled with water and direct sunlight. When building the container, it needs to be watertight. Your other option is to do an inground hydro pool. If you live in a climate where you need not worry about winter and frost, you could dig a large hole that would make a small lake or water trough. It needs to be made to prevent water from leaking into the surrounding ground. It need not be deep. Only about six inches to a foot of water is necessary for most plants. A pump is necessary to circulate the water every few hours. The garden requires at least 10 hours of sunlight or plant lights. Pack the starter seeds in a growing medium and let nature take its course.

For smaller water gardens, you can use containers and a large fish tank. A pump will circulate the water, while you can have water pumped into a variety of containers from the holding tank. The fish tank would be the holding tank. On a cycle, the water circulates goes through a filter and goes back into the holding tank. With individual containers and a holding tank, you can the plants individually rather than having one giant pond filled with a variety of roots and plants.

Water Catchment Systems

Whether you choose a hydro-garden or a raised bed design, you still need water to help you grow your food. Living off-grid usually means a well and septic for your needs. A well might produce as much water as you need without ever drying up. However, sometimes a well needs time to refill, which could mean not enough water for showers, washing dishes, and other basic needs, let alone watering your garden.

The amount of rain you receive in your area per year also determines how moist the soil stays. Water catchment is one way to ensure the garden has enough moisture, during dry seasons especially. A 500 sq ft roof can capture about 300 gallons of water during an inch of rainfall. For a small roof, that is a good start toward ensuring your garden is watered thoroughly when rain is lacking. There are two ways to make water catchment work.

> 1. Let the rain roll off the roof, through gutters and let it run downhill into your garden.

> 2. Use gutters and a rain-collecting device, such as a barrel, that you can then transport the water to the garden when you need it.

There are other water catchment options. Just sitting barrels out in the field can ensure you collect rainwater as it falls. Although many find a roof, even a shed roof is better for catching more water. With hydro systems, the pond or holding tank you have can be the catchment device making it easier to keep your garden growing.

Irrigation

With catchment systems that include a barrel for ground or raised bed gardens, you will want a pump that will help you distribute the water throughout the garden. You can use an elaborate irrigation system, with pipes running in the ground and sprinkler heads that pump the water onto the leaves. There are also inground choices that do not have sprinklers and simply feed the roots.

Irrigation has been around for centuries, so living off-grid with an old-fashioned pipe system is possible and quite reliable.

Drip irrigation requires 10 to 30 psi (water pressure), but even 8 psi can be enough. You do not want too much pressure because the leaves and plants could become damaged.

You may also consider using a river or stream nearby to help you get water to the garden via a pump. If you live in an area without rain or where water catchment is not as great, but you have a river or stream, then pumping water from the stream, through pipes, and into your garden is another way to ensure you are getting the proper amount of water.

Pest Control

Moving on to pest protection, gardens are always a "hotbed" of desire for various pests. Deer, boars, bears, and smaller pests like insects and groundhogs can be very detrimental to your garden. Living off-grid, you probably want to keep your food organic, given it is healthier, which is why it pays to ensure proper pest control is available.

1. Solar Fencing

Solar fencing is for the larger rodents and pests that might try to eat your garden. Any fence you put up around your garden should be buried at least a foot into the ground. The deeper you go, the easier it is to keep rodents out. Rabbits, foxes, and raccoons won't dig deeper than 12 inches. Groundhogs and other underground dwellers are a different matter. They can burrow under the fence and attack from the bottom. Solar fences are also used to keep deer and other grazing animals out of the garden.

Solar energy powers the fence to give a small electrical jolt to the larger pests. You want a tightly woven fence, such as chicken wire, to help keep out the medium-sized pests. By using electrical wire and a solar panel, you can electrify the fence.

Deer and elk can jump fences, and you may not want to make it too high. It sounds cruel, but baiting the wire with things deer like to eat helps them realize they do not want to touch the fence. This can backfire. Deer and elk are such great jumpers they may not have to even touch the fence to get over it if it isn't built high enough.

2. Natural Insecticides

Keeping small insects out of your garden can sometimes be harder than the large pests. You don't want to use pesticides that will harm you and your family, so learning about the 8 natural insecticides will help you come up with a concoction that works for the pests you will encounter.

1. **Soapy Water** - 5 tablespoons of dish soap in 4 cups of water kills aphids and mites.

2. **Neem Oil** - Readily available as an essential oil—neem is derived from trees that grow in India. It is an anti-fungicide, kills scale, mites, and aphids, and other insects.

3. **Pyrethrum Spray** - Created from chrysanthemum flowers, the powder should be mixed with water to help stop flying insects.

4. **Garlic** - A vegetable not only helpful for cooking, but also stops bugs. Be careful with growing garlic with some of your food plants though since they can compete.

5. **Beer** - And you thought beer was just for drinking. Beer is made with yeast, barley, and other grains, which you might think attracts bugs. However, due to the alcohol and other ingredients, insects are more attracted to a saucer of beer than they are your plant root, leaves, fruit, and vegetables.

6. **Pepper** – Some people use pepper spray using a red pepper, dish soap, and water. Any pepper, including paprika, helps stop spider mites and other aphids. Take 2 tablespoons of pepper, 6 drops of dish soap, and a gallon of water, mix it, and put it in a spray bottle.

7. **Herbal Sprays** – Thyme, sage, rosemary, rue, lavender, or mint crushed and soaked in water can make an herbal insecticide. Even putting these herbs in your garden and growing them for recipes can help your pest control efforts.

8. **Nicotine** – Using 1 cup of dried tobacco leaves and a gallon of water ensures you can keep leaf-chewing bugs off your plants.

The nicety about using these different methods is that they are safe. With a thorough washing of your plants, even dish soap is less harmful than the commercial pesticides sold for gardening.

Composting

How will you feed your garden? Soil may have enough nutrients for one year of growth, but after that, you will need to add some healthy food to your garden. Composting is one way to keep your garden growing long after the original soil has lost its nutrients.

An area of your land should be set aside for a compost pile or three. Three is best—one you use, another ready for the next year, and a third you add to throughout the year.

The benefits of composting include:

1. Cost-effective due to you and your animals contributing to the pile.

2. It is a natural way to ensure your garden soil has the appropriate nutrients.

3. Your compost ensures the healthy growth of plants.

4. Composting reduces the pressure on landfills.

5. Composting also adds to water conservation because the soil can retain more water.

6. Composted materials have natural fertilizer.

Compost Ingredients

1. Straw

2. Leaves

3. Sawdust

4. Newspaper

5. Wood chips

6. Grass clippings

7. Vegetable peels

8. Tea leaves

9. Eggshells

10. Coffee grounds

11. Weeds

12. Fruit

13. Wood ashes

14. Garden residue

15. Waste

The size of your garden determines if buying or making a compost bin is better than having a large pile of ingredients taking up a 100 sq ft space. The key to any compost pile is to ensure it has plenty of oxygen and moisture to work properly. The matter needs to be broken down, so it works as a compost. Any compost you use should have the moisture level of a wrung-out sponge. It is better to go for too dry than too wet. Too wet can lead to issues of mold.

Types of Bins

1. **Tierra Garden** – Created from recycled plastic, this 80-gallon bin helps you compost kitchen and garden waste. It is prebuilt to help you compost, while also keeping it tidy and critter-free.

2. **Wooden Box** – Wooden building boxes can be another way to create piles and keep them tidy. Old pallets, discarded building materials, and logs have been used to make composting bins. The one thing you want to take care of is to use untreated and unpainted wood. You don't want to add chemicals to your compost through the wood you use to build the bin. With a compost pile, the bottom will rot before the top, so shallow bins you rotate the use of will ensure your compost remains usable.

3. **Countertop Composter** – If you have a smaller garden, a countertop composter might work better. These bins are expensive because they are made with recycled plastic and contain compostable bags to keep the process simple.

The size of the bin determines how long it will take for your compost pile to be ready to use. The process can take as little as 3 months in small pins and up to 2 years if you just keep piling things on top of the original pile.

Aeration is necessary for the oxidation of compost materials. You do not have to use a bin, especially if you want to turn the pile so oxidation occurs and helps the microorganisms break down the material. Aerating compost every few days is the best routine.

Watering your pile depends on how humid and warm your climate is. A dryer climate will require more watering when in the direct sun versus a moist climate that may never allow the pile to dry. A soggy pile will rot and not go through the proper aeration process. Instead, it will mold, and if you use the compost, it will cause mold to grow on your plants. Vegetable and fruit mold differ from the mold that grows from being too wet. If you have aged fruit and vegetables with mold, those mold cells will help with the decomposition of your compost materials.

Uric acid or urea is helpful in compost piles. Get into a routine of dumping your waste from a composting toilet in the morning to allow it to move from the top of the pile and soak into the material.

Chapter 10: Raising Livestock for Food

Don't read if you are a vegetarian! Only kidding... You are reading this chapter because you want to know how to raise different livestock for food on your off-grid land. Raising livestock can be an expensive concept because it requires having enough space, includes slaughtering fees, veterinary care, maintenance, guarding, and raising enough to feed your family. If you are a beginner homesteader, start with chickens. They are low cost compared to cattle, goats, and pigs. Throughout this section, you will gain knowledge on raising livestock and slaughtering, particularly the crucial factors of local ordinances and costs for using controlled slaughterhouses.

Raising Chickens

Over 500 breeds of chickens exist; however, there are perhaps a dozen you will be interested in adding to your homestead. Ameraucana chickens are some of the best egg-layers you can ever have, plus the give you blue-shelled eggs.

Chicken Breeds

1. **Ameraucana Chickens:** A medium-sized bird that produces large blue eggs. Sometimes Ameraucana can lay more than one egg in a day and typically provide 250 eggs per year. Ameraucana have dark feathers, often with some silver and blue.

2. **Australorp:** A black chicken with a large comb, the Australorp was brought to the US from England, although they are an Australian bird. The Australorp produces at least 250 brown eggs per year and is even known to lay 364 eggs without artificial lighting.

3. **Bielefelder:** This is a dual-purpose chicken because it is a great egg-layer, providing at least 230 brown eggs per year, but is also a 10 to 12-pound fowl. Once it has reached the end of its egg-laying years, it can become food on your table to make room for new, younger chickens.

4. **Orpington:** Another English breed, it was breed by William Cook for eggs and meat. Orpington chickens can be orange-red in feather color, but also black, white, buff, jubilee, and spangled. Orpingtons are known for laying at least 200 brown eggs per year. They are also the leading chicken for those who want a pet versus a meal.

5. **Plymouth Rock:** A mostly black chicken with spangled white plumage; the Plymouth Rock is an all-American bird. It was created for its eggs and meat, plus its ability to adapt to hardy conditions. The Plymouth Rock is a brooder, with docile behavior. The Plymouth Rock also produces brown eggs.

6. **Rhode Island Red:** Another large bird, the red-feathered species, offers five to seven brown eggs per week and then becomes a great fowl for the dinner table.

7. **Cinnamon Queen**: The cinnamon queen is known for its brown eggs, but they can also produce white eggs since they are a cross between the Rhode Island red breed and white breed. This is a breed to have if you want egg laying to start quickly. Pullets lay 250 to 300 eggs per year, and typically start a week or two before other pullets born simultaneously.

8. **Barbezieux:** A French breed, the Barbezieux weighs 9 to 12 pounds in adulthood. While not breed specifically for egg laying, this breed does provide a significant amount of white eggs. However, it is the firm and distinctive flavor of the meat that farmers appreciate.

9. **Cornish Chicken**: Also known as Cornish Game Hens, you hear about these birds being a small, singular meal for people. The hens are usually harvested at one pound, considering this is when the meat is most tender.

10. **New Hampshire Chicken**: A red-feathered bird, the New Hampshire chicken can be dual purpose. It's a happy free-range bird that can produce a significant amount of eggs before you add it to a table meal.

Remember; there are plenty of breeds to add to your farm. The above ten were chosen based on knowledge of raising chickens and personal preferences.

Chicken Life Cycle

All chickens are hatched from a fertilized egg. You can buy fertilized eggs and help your chicks hatch, or you can purchase chicks to begin your homestead. An egg hatches after about 21 days. When a chicken hatches, it will have a wet "down" feather, which dries quickly and turns them into a fluffy adorable chick.

Chicks need to be in a brooder. A brooder is an indoor space with a heat lamp. In the first couple of weeks, clean the brooder each day and supply a baby with chicken feed. After five days, a chicken will show some real feathers, and by day 12, you will see a significant breed indication. By 18 days, the chicken will have most of its regular feathers with very few downy ones left. After a month, more breed characteristics appear.

Young chickens that have reached egg-laying status are called pullets. At 18 weeks, most pullets are ready to lay their first eggs. Healthy well cared for chickens will lay eggs for three to five years. The number of nutrients, warmth, and care you provide your chickens will determine how frequently they lay eggs. Their breed also indicates if you will see five to seven per week or a cycle of one egg every other day.

Chickens molt annually to remove old feathers and grow new ones, which can cause a slowdown in their egg production during the molting process.

Unless you intend on allowing a rooster to fertilize some eggs for new chickens, you probably want to buy already hatched chicks to avoid getting a rooster in your pen.

Many cities and even counties have laws against roosters. For example, one county states that a person can raise up to six chickens for egg laying or meat but cannot have a rooster. If you are homesteading and have a commercial license for livestock, the rules can differ. You may be able to have a rooster on a "farm," despite county or city codes. Research this aspect based on your city, county, and state.

Chicken Coops

Whether or not you will raise free-range chickens, you still need to have a chicken coop. A coop is a place for chickens to brood, ensuring you get the egg production you desire. Some chickens will not care if there is a nice pile of bedding to roost in and then sit on their eggs. These chickens tend to drop an egg wherever they are at the moment. Others are great brooders, so they will dig into the dirt or pile bedding up and then lay their egg.

A coop is also necessary for the protection in the dark hours. Coyotes and wolves are just two predators that enjoy a meal of fresh chicken. By providing an indoor space, you keep your birds safe and less stressed.

The coop needs to have a 1 sq ft space per chicken. It should have a vent, a ramp to get inside, and a larger door to help you access the floor for cleaning. A window and even a lamp are a clever idea to help keep your birds happy, in the sun, and warm.

Any coop you build or buy should be airtight. Think of it as a mini home where your chickens can winter warmly without a heat lamp. The cost of creating a coop can be as little as $100.00 when you use recycled materials, nails, and tools. Buying a coop can range from $100 up to $5,000. The larger and more elaborate the enclosure, the more it will cost. Fencing

Even free-range chickens should be contained within your homestead land. The idea of free-range is that the chicken can roam and eat the grass and seeds that naturally occur on your land, rather than being cooped up in a small building or small fenced paddock. Free-range chickens can mingle with other livestock.

Any fencing you have should be buried at least a foot deep. It should be "chicken wire," which is about a centimeter in size per square hole. This prevents smaller pests like raccoons from getting in and stealing eggs or harming your chickens.

If you have a lot of pests in the area, it can be a clever idea to have a wire roof over the fenced area. There are rolling fences that let your chickens graze on your open land without being vulnerable to predators.

Food and Water

There should always be water available for your chickens. There are a variety of water containers from automatic waters to a simple bowl on the floor. A wall container is better because the water remains clean versus a bowl on the floor that could be used for bathing, plus excrement. Some chickens use bedding to bury the bowl of water. However, if you live in a cold climate, you may need one of the floor bowls with a heater inside. The heater will keep the water from freezing during the cold months.

As for food, you will need to begin with chick feed if you buy baby chickens. Chick feed has more nutrients that help chickens grow into healthy egg-layers. Once a chicken starts to lay eggs or reaches 16 weeks, it is time to switch to adult food. Adult chicken food contains oyster shells for calcium, plus other helpful nutrients for a happy, healthy bird. Calcium is imperative for a strong shell. If you do not have a food high in calcium, then the shells will be a weak and possibly lead to too many lost eggs.

If you buy food, Layena by Purina is a good, clean brand. It is affordable with the added plus it is high in oyster shell calcium. Chickens can also be fed table scraps. Free-range chickens will eat insects, pull things from your garden, and get nutrients from the ground. You can supply eggs, rice, vegetables, and fruits for your chickens.

However, realize your chickens are birds of routine. If you feed a store-bought food like Layena crumbles, it may be impossible to change their diet. In one case where the chickens ran out of food and had only pellet food, it was discovered that the chickens wouldn't eat the pellets because it was too much work. These same spoiled chickens also avoided any homemade food.

If You Want Non-GMO, Totally Organic Food, Consider the Following:

1. Alfalfa meal for protein

2. Corn

3. Peas for protein

4. Wheat

5. Oats or Barley (less than 15 percent of the other ingredients)

6. Aragonite, limestone, oyster shell

7. Grit

8. Salt

9. Crab meal

10. Flaxseed

11. Kelp

12. Fish meal

13. Cultured yeast

A mixture of the above ingredients ensures a healthy diet with no commercial made food. Note, the amounts are not provided because you need to make batches based on how many chickens you have and whether you have storage for a few days of food or need to make a mix each day.

Raising chickens for eggs differs from raising them for meat. A meat chicken grows rapidly and eat a lot of protein to grow meaty for you to enjoy. Although you can allow your hens to produce for a couple of years and then slaughter them for a meal, generally the younger the chicken is the better the meat will be. If you buy chicken breeds intended for meat, they are supposed to be slaughtered within a year, or their heart may give out. These are the chickens genetically breed to rapidly increase in size.

Chickens may be the best livestock to start your farm with, but there are six animals you can raise on just a quarter of an acre. If you have a full acre, you can add in more livestock.

5 More Livestock to Keep

The livestock in this category is like the ease of care and space to chickens. If you want small animals for meat or eggs, then the following will be beneficial to your homestead:

1. **Ducks:** Much like chickens, you do not need a huge space to raise ducks. Male ducks do not crow, so they are also allowed in neighborhoods that might be against roosters. Ducks need 4 square feet of coop space per bird and plenty of running around space. Like chickens, you can raise ducks for both eggs and food.

2. **Dwarf Goats:** Goats are multipurpose animals because they can offer meat and milk. One goat requires 5 to 6 sq ft of space, plus 20 feet of grazing area. If you do not want to tackle raising cows for milk or beef, then a goat may be the next best option.

3. **Sheep:** Another multipurpose animal that provides meat, milk, and fiber is sheep. Shearing sheep for wool to spin and make clothes from is an added benefit to raising sheep on your farm. Unfortunately, sheep are more difficult to milk due to their instinctual behavior against predators. A grown sheep needs 12 to 16 sq ft of living area plus 16 to 25 sq ft of grazing area.

4. **Quail:** Quail is a small fowl that requires less space than ducks. Unfortunately, they can be on the noisy side, which is not good for city living. Off-grid dwelling makes a perfect place to raise quail for their eggs and meat. They eat

less and require less bedding, so you can raise more quail per sq ft than chicken or ducks, for less money. However, you get less meat.

5. **Rabbits:** I cannot think of eating Thumper, but rabbits have been food for centuries because they are easy to raise. Rabbits can be destructive creatures. Although many consider it a myth, rabbits can crawl into cars and other tight spaces and chew wires. Rabbits are good for fertilizer, meat, and angora wool. They eat little, and you can feed them natural grasses to keep costs down. You will need to provide at least 2 sq ft per rabbit and up to 5 sq ft for a 12-pound rabbit.

Cattle

Cows or cattle need to be in their own section because a lot goes into raising milk cows and beef cows. First, a beef cow takes at least a year to raise, where you help it grow from a 500-pound calf to a 1,000 or more-pound cow. It takes at least 2 acres to feed a calf and raise it to a cow. One beef cow can feed a five-person family for one year, as long as you have other options like chicken, pork, and fruits and vegetables.

Cows do graze on grass, but you also need to feed additional nutrients. Many cattle owners will provide corn to help their cows gain weight. You also need to provide bales of hay. The formula for figuring out how much hay is to consider a cow needs 3 pounds of hay per 100 pounds of weight. So, a calf that weighs 250 pounds needs about 7 pounds of hay per day. If you have a lot of grassland, your cows may get enough grass by grazing.

The vet fees for taking care of your cattle can also be extensive. Depending on where you live and how close the vet is, you can expect to pay at least $45 for the visit and an additional $40 to $50 for any vaccinations that may be needed. In California, the costs can be higher at a whopping $90 per non-emergency visit. Note: these costs are for a regular visit to ensure your cattle are healthy. Fees for birthing help, illness, or emergency can increase the cost to well over $150.

Angus and Hereford cattle are the best for meat, but there are other choices you can buy.

Pigs

Pigs are another way to raise meat on your farm. A 200-pound pig can provide enough pork, ham, and bacon for a family of four, while two pigs are best for up to eight people. You can have a farm full of pigs that produce baby pigs you can raise for slaughter, or you can continue to buy piglets you raise until it reaches 225 to 325 pounds.

Buying a pig in March or April helps you get the pig ready within a year for slaughter.

You will need to pay vet fees for the inoculation of various diseases. A pig should be inoculated for cholera at eight weeks. You also want to ensure you buy a pig that has already been through a deworming regime. Pigs can be expensive, not only for the grain you buy and the vet visits, but for their full care.

Your pig will need a pen and a house. A simple structure 8 x 6 x 5 feet will suffice.

At 200 days, your pig should be 325 pounds.

There is no specific breed of pig that is better than another when it comes to meat.

Slaughtering Livestock

Being self-sufficient is one thing but having the time and knowledge to slaughter your livestock is completely different. It takes years and more than a short section in a homesteading book to discuss the best methods of slaughter.

1. Make sure the laws of your state allow for the slaughter of livestock. If you intend to do it yourself, the laws may restrict you to where you must find a creditable slaughterhouse.

2. Research slaughterhouses and prices.

3. Assess their methods and the cleanliness of the location.

4. Make sure the process is humane.

5. Seek a professional who can cut the meat correctly and use all parts of the animal.

6. Most places will help provide bacon, which can be hard to make on your own if you slaughter your own animals as a novice.

Protecting Your Livestock

Beyond fencing the grazing land, you want to have a guard dog. Several breeds of dogs are bred to be herders and protectors of livestock. A dog can deter a larger predator by being around and on watch. Any livestock dog will make an excellent pet too, but you want to encourage them to bark only at danger. Not all dogs will make the best guards due to personality differences. Some breeds meant for guarding can also produce a lazy dog, while others in the same litter can be the best guards you will ever see.

It is best to seek a breeder for a non-mixed breed and to be clear in what you want from your puppy. Starting with a puppy ensures that you raise him or her for what you wish to see in behavior.

Chapter 11: Preserving Your Food

Did you grow too many tomatoes? Or maybe it was too many blueberries? What will you do since you don't want to waste the food, but you and your family cannot eat it fast enough? Preserving food solves keeping your items fresh and edible throughout the year or years you take to eat everything. As a homesteader, who wishes to be self-sufficient, learning how to can, ferment, pickle, dry, and freeze various foods guarantees self-sufficiency. Other food preservation techniques discussed are cold storage, using a root cellar, and salt curing and smoking to preserve meats.

Drying

Drying requires a food dehydrator, sun, or shade. Shade or adiabatic drying is a process without heat. Solar or sun drying helps the fruit or vegetables dry in the sun, where the container captures the heat. Apricots, grapes, and tomatoes are the most common foods sun-dried.

A food dehydrator uses air and heat. You can also achieve the same concept by using a warm oven, set at 165 degrees F. Using a food dehydrator is safer because it has a timed setting for the foods you might wish to dry, including jerky. Making jerky in a warm oven can go wrong, causing microorganisms to grow. You want the moisture to evaporate quickly, so there is no possibility of bacteria surviving the process.

Once the food is properly dehydrated, it can be stored up to 12 months, depending on the meat, fruits, or vegetables.

Canning

Canning helps you seal the food to prevent it from turning. Jelly made from fruit is one way of canning. You may also can tomatoes or turn them into spaghetti sauce before sealing the container against oxidation.

The process of canning makes food self-stable if the food is in an airtight, vacuum-sealed condition. It is important for all foods canned to reach a temperature of 250 degrees F during the process to render enzymes inactive and microorganisms dead. As the food cools, in the vacuum-sealed container, it cannot grow new bacteria.

When canning, it is imperative you do so using appropriate guidelines and not a random recipe or method found online. Home-canned foods contain a higher risk of botulism.

Canning your food requires you to leave a headspace, so the proper vacuum forms. Typically, the liquid should stop at the neck of the canning jar, so there is an inch of headspace between the liquid and the lid.

Use any canned foods you prepare within a year. If you feel the seal on any can have been compromised, throw the contents away.

Boiling Water Canners

Boiling water canners are usually porcelain or aluminum. They have perforated racks with fitted lids that help you boil the water to the correct temperature. Visit **https://nchfp.uga.edu/** for proper canning instructions. It is a USDA backed PDF that tells you how to go through the canning process for the utmost safety.

You can also buy a canning machine that takes all the guesswork out of preserving your food. You simply put in the food, and the machine does the rest of the work.

Root Cellar

A root cellar helps you store carrots, onions, potatoes, and other root foods in a dark, underground place. Apples and tomatoes can also be put in a root cellar.

Freezing

Freezing prevents something from spoiling until you can use it. You can freeze almost any fruit, vegetable, or meat. You want to keep the food completely frozen and avoid anything that may have thawed and then refrozen. Freezing will not kill microbes like bacteria or parasites, which is why you need to ensure your livestock is vaccinated and that anything you grow is parasite-free. Freezing will not destroy any of the nutrients in the food; however, left too long in the freezer, things can become freezer burned. Although the meat is safe, it's tougher meat so most people cut off the dried bits. The types of packaging you choose determines how long something can remain in a freezer without deteriorating. As soon as you slaughter your livestock and package it, you will want to freeze it to ensure it retains the proper quality. A freezer should be at 0 degrees Fahrenheit for the proper temperature of the items inside.

When you thaw the food, do so in the refrigerator or in chilly water. It is better to thaw it in the fridge overnight rather than leaving it on the counter or in water that will warm. It is also possible to use a crockpot and thaw the meat as it is cooking for 8 hours.

Meat, uncooked, will last in a freezer for four to twelve months. Bacon, sausage, ham, hotdogs, and lunchmeat are good for up to two months. Casseroles, gravy, poultry, cooked meat, soups, and stews are good up to four months. Egg whites, uncooked chicken, and wild game can last for a year.

Pickling

Pickling entails putting something in a vinegar liquid to keep them longer. Pickles made with cucumbers through the pickling process are the most common. However, cabbage, radishes, jalapenos, banana peppers, and other vegetables can be pickled to preserve them for use later.

Salting

Salting is a method of preservation used for centuries. One of the most typical meats preserved through salting is ham. When ham is salted it can last for more months than if it was left uncooked.

You do not have to cook the ham for it to become dry-cured and remain stable for several months on the shelf. However, you do need to make sure you have used enough of the salt and other ingredients to successfully remove all moisture from the ham. Dry-cured hams take a few weeks to a year to age properly. Six months is the typical amount of time a dry-cured ham requires for the correct taste and softness so it's ready to eat.

Smoking

Smoking, which uses a wood smoker, helps you cook the meat with flavor and ensure it lasts longer than if it was frozen uncooked. Smoking can be done in a wood smoker or over a fire that slowly smokes the meat. You want wood chips that will provide a flavor to the meat. Check for the correct temperature of smoked meat since ham, chicken, and beef will need different temperatures before they are considered done. Smoking is like salting because you are dehydrating the meat to prevent bacteria from growing. The reason you might wish to use smoking over salting is the salt. Too much salt is never good for the diet, so using hickory, oak, cherry, maple, or applewood chips ensures you have a nice flavor without too much salt. Smoking generally requires eight hours, if not a full 24 hours. The longer you smoke the meat, such as 48 hours, the longer you can keep the meat.

Chapter 12: Making Money From Your Homestead

Once your homestead is up and running, it is time to use your resourcefulness in gardening, farming, animal husbandry, beekeeping, and off-grid living to turn a profit and supplement your needs. There are a variety of ways you can earn money by selling your oversupply of vegetables and fruit, and selling other items you raise or make. Farmer's markets are a wonderful place to offload extra food, and town festivals give you a chance to sell things like homemade soap.

There are at least 50 ways you can make money from your homestead.

Money from Animals and Insects

The following will discuss how you can make money from animals and insects you raise on your homestead.

1. Bees

Becoming a beekeeper helps you make money because of the honey you can collect from honeybees. In a time when honeybee populations are endangered, it is more important for homesteaders to do their part in raising colonies and enjoying the benefits that come from bees.

- Start small, such as one honeycomb.

- Seek a professional's advice on how to approach the bees and collect honey so you may avoid endangering your life or the bees.

- Collect the honey for yourself and sell the extra.

2. Breed Livestock

Homesteaders have not only raised cattle and other livestock for meals, but also for sale. You can breed livestock to sell to other homesteaders just getting a start, or you can sell the beef you gain from raising livestock.

There are three ways you can make money from livestock.

- Breed them for sale.

- Breed them for slaughter and sell the meat.

- Become a livestock consultant. By enticing people to your homestead to learn from you and gain advice, you can make money to help feed your livestock.

- Breed Livestock Guard Dogs.

If you have a great pair of guard dogs that help protect your animals, there is no reason you can't breed them and sell their offspring. Finding the correct personality for watching over livestock is not always easy. Breeders create dogs for many reasons, including pets. Finding a breeder that provides dogs for guarding livestock are fewer.

Most purebred dog breeders can get a couple of hundred dollars or more for their dogs.

3. Making Products to Sell

Making your own products and selling them is also time-honored in the homesteading life. Soap is one of the biggest ways to make money online and at art fairs. Castile, goat milk, olive, coconut, shea butter, and lye-based soaps are something our ancestors had to make. Typically, the original soap was made with lye, but now you can use things like goats' milk by raising goats.

One of the things that are lacking in the world is homemade soap without additive fragrance or perfume. Using natural oils, such as mint from your mint plants, is much better than any fragrance you could buy and use. A huge movement is underway right now to use natural, organic soaps that lack chemicals but use natural oils to ensure healthy skin.

Soap is just the beginning. You can move on to shampoo, conditioner, lotion, lip balm, and other toiletries.

With vegetables like avocado and herbs such as mint, you can add ingredients designed to help the hair and skin become healthier. Several herbs and plants you grow can have natural remedies for dry skin, aging, and even pain.

Consider making products with antioxidants, anti-inflammatory, and anti-bacterial properties as these will also sell due to the natural ingredients.

4. Firewood

Cutting down healthy trees is obviously not something you want to do, and depending on where you put your homestead, you may not have a lot of trees. However, for those who live near forests like Colorado, fire mitigation is essential to keeping a homestead safe, and it provides plenty of firewood for you and potentially others.

Anytime you clear land of trees or conduct fire mitigation, there is nothing wrong with selling bundles of wood to those who needed.

If you do not want to cut the firewood, you can also offer a lesser cost for a person to come and chop firewood for their personal use. If you already have downed trees, then paying someone to haul away the trees for firewood can get you a little extra funding.

5. Growing and Selling Vegetables/Fruit

Anytime you have an overabundance of veggies and fruit, even after canning some to save for the year, you can get a booth at a farmer's market and sell your goods. You want to choose the best items to bring and have comparable costs to those selling around you.

Another option is to get a license that allows you to sell to a restaurant. Restaurants in certain areas are always happy to pay a reasonable price for homegrown mushrooms.

6. Freelance Work

As a homesteader, there are a lot of things you have learned or perhaps already knew. You can use these technical skills to make money. Writers use sites like Upwork to connect with clients looking for books about homesteading. If you don't want to share your knowledge for a few cents per word, you can always author a book and self-publish on Amazon and other online sources, like Bookshop.org. You will need marketing skills, but you can make money from your knowledge.

Offering your farming skills, woodcutting, or property for livestock are other ways you can freelance for profit.

7. Making Clothing

Remember those rabbits with Angora wool and the sheep and goats with their fur? For anyone with the skills, sewing and knitting clothing, towels, and blankets are ways to make money. People love homemade items of quality and will be willing to pay for them. Like the soaps, you can make your clothing and other items to sell at art festivals or create an online website where you sell whatever you make.

8. Candles

Candles are other homemade commodities that people will purchase. Plus, making candles is actually easy. You need wax, wicks, and a container. You will also need a pot to melt the wax in and color or oil to make the candles beautiful. A mason jar is a good container because it is clear. It also provides a base for the person to continue burning the candle until finally it is gone.

As with soaps, finding non-perfume or fragrance laden candles is hard. Essential oils are a better option or no scent at all. In fact, there is an entire untapped population who would buy candles if they were just pretty in color and lacked any oil or scent. This population has an allergy to fragrance-based smoke, and there are more of us out there than you might imagine.

9. Bread

Making bread from the yeast and other ingredients is possible. French, peasant, and Amish bread are just a few options. Here is one recipe you might want to try:

Amish Bread

2/3 cup sugar

2 cups water (warm)

1 ½ TBLs yeast

¼ cup oil

5 to 6 cups flour (all-purpose)

As a rising bread recipe, you will need at least four hours for the bread to double in size. Then, you will need to knead it, and let it rise a second time before baking it at 350 degrees F.

Any bread recipe you find will add to your table, but also to your pocketbook because you can make several loaves of bread to sell at farmer's markets alongside the fruits and vegetables you grow.

10. Cheese

Goat cheese is another way to earn money, but there are also plenty of other types of cheeses you can make and sell at markets.

11. Selling Eggs

Selling your chicken quail and duck eggs are another three ways you can make money from your farm. By selling the eggs, you also raise money for the feed it takes to keep your flock healthy and productive.

People will pay at least $4.00 per dozen for fresh, free-range eggs.

12. Starting an Orchard

Depending on where you live, it may be possible to grow grapes, apples, and other fruit trees. You not only benefit from the fruits grown, but you can also open your orchard to those who may wish to come pick their own fruits. As a seasonal pick location, it is possible to make more money.

13. Raising Bait

Do you live in an area with a lot of fishermen? Do they need bait? There is no reason you cannot start a worm or bait farm that would make you money. Worms are just one type of bait you might raise. You can also raise fish and use their eggs for bait to sell to local fishermen.

14. Renting Your Property

Do you have some land you could build a little cabin on? Perhaps you want to take a little break from the homestead? By putting your home on Airbnb you can rent it out and make a little money. If you have an area for camping or RV storage, you can also make a little money by collecting a small renter's fee.

There are at least 50 ways to make money from your property. While only 15 have been mentioned, just think of what you can do with your skills. Perhaps you made your own furniture, or do you know how to make cabinets? Maybe you are great with pickles, jams, jellies, and syrups. Any skill you must feed your family and keep them sheltered can turn into more than a hobby and a way of life—it can become a method to make money from your land and skills.

Conclusion

Congratulations on reaching the end of this guidebook. You have it to refer to as many times as you need to become successful in your off grid living choice. Whenever you have doubts or questions, let this book be your answer to moving from novice to expert homesteader.

You knew the general definition of off grid living, and now you understand that you can choose how extreme you are and whether you raise livestock or take advantage of a nearby farm.

You learned you don't have to be off grid without proper water, electricity, and sewage disposal.

It is up to you to take advantage of the different chapters and set up a homestead that will work for you.

Remember, your commitment to being off grid, even with the Internet, will determine your success. Start small and make your way towards larger goals such as raising chickens before getting a cow to milk or raise for beef.

Your first step, now that this book is complete, is to either find the property for your homestead or to research where you can live comfortably. Do not forget your due diligence in researching the city, county, and state laws so you may have a successful homestead and perhaps a business one day.

Living off the land like our ancestors is not only rewarding, but an essential means of survival.

Part 2: Raised Bed Gardening

The Backyard Gardening Guide to an Organic Vegetable Garden and the Best Way to Grow Herbs, Fruit Trees, and Flowers in Raised Beds

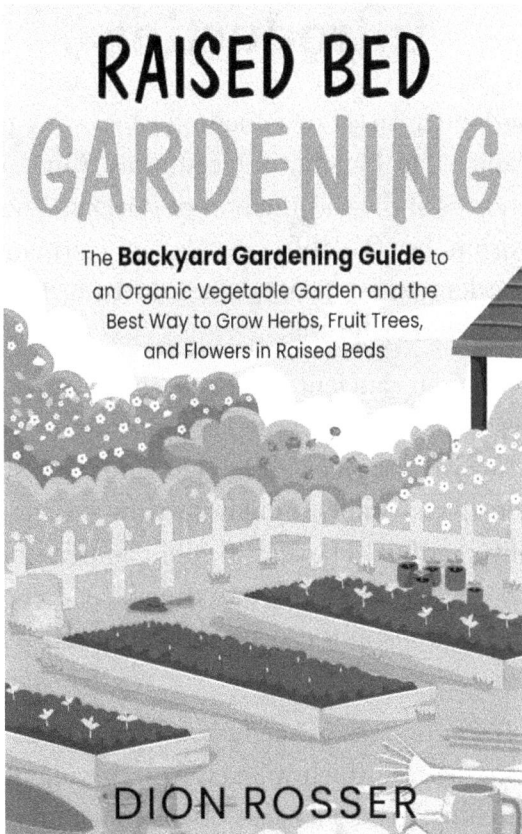

RAISED BED GARDENING

The **Backyard Gardening Guide** to an Organic Vegetable Garden and the Best Way to Grow Herbs, Fruit Trees, and Flowers in Raised Beds

DION ROSSER

Introduction

Raised bed gardening isn't a new method of growing plants; people have been reaping the benefits for centuries. One of the earliest recorded references about this method of gardening was in Liber de Cultura Hortorum by Walahfrid Strabo, a German Benedictine who praised gardens with poems written in Latin hexameters, also called the little garden or hortulus.

Although it is an ancient horticultural concept, raised bed gardening has suddenly become popular again, especially in overpopulated urban regions with limited gardening space. It is the perfect solution if you want to cultivate your own food or simply own a decorative garden. They are flexible and can be constructed in and for tiny spaces, including paved and concrete areas. If the current garden has no soil, it can be supplemented with any soil of preference.

This book is a comprehensive guide with easy-to-follow chapters. It begins with the basics, which will take you through the concept of raised bed gardening, the advantages of engaging in the practice, and everything you need to know before getting started. Reading further, you will be educated on the materials and tools of the trade, especially the materials from which you can construct your raised beds.

It also advises on the preparation and construction of your garden, including positioning to maximize sunlight, which is especially important, as you will discover soon enough. If the garden will be constructed in frost pockets or shade, this book is your go-to guide for suitable plant suggestions. You will also be provided with tips to help you design your garden layout to create textural and thrilling color combos. As you read further, you will discover various ways to identify and handle common plant diseases and pests.

After internalizing the basics, we'll look at the different raised bed gardens you can construct, including everything from purchasing already constructed products to making your own bed from scratch with the use of recycled materials.

You'll also learn about the less popular raised bed garden options such as the keyhole garden bed, which is an example of a sustainable raised bed that is mostly found in Africa. It features a universal compost heap and is built from any free materials that surround the area, like rocks excavated from poor soil. Another type of less popular raised bed is the hügelkultur, which originates from Northern Europe. It involves constructing beds over rotting tree stumps and timber, which slowly decompose to produce nutrients and improve the soil. One of the most attractive features of raised bed gardening is that everything from proper construction to actual planting can be done by yourself.

This book is rich in suggestions for what you can create or build from the comfort of your own home and in the privacy of your own space. You will discover a variety of planting schemes to choose from according to your personal circumstances and preferences. These varieties range from mini orchards to vegetable plots and even Japanese gardens. There are also options for people without an actual garden, like green roof space and window boxes.

I hope you will be inspired by this book to delve into the world of cultivating plants in raised beds. Whether you own a little courtyard, a grand walled garden, or a community garden, this method of gardening will improve and enhance your outdoor space organically and aesthetically. Once you get started, the ease of gardening will ensure that you never look back.

Chapter One: What is Raised Bed Gardening?

Gardening can be difficult to do, especially if you're struggling with a physical disability, you have issues with mobility, or you simply choose to enjoy your later years in life without having aches and pains in your legs and back. Constantly bending forward can make gardening difficult for some people, and this is where raised beds come into the picture, with a host of other advantages.

Definition

Raised bed gardening is a method of farming that provides areas for cultivation on rocky ground and raises the garden to a level that is more comfortable for the gardener. This method of gardening can help to minimize insect or pest attacks and protects the plants from excessive flooding when it rains heavily. It also works to keep the soil warm in early spring.

Think of a raised bed as a large container sitting on the ground with no base. It is a simple structure that is elegant and can be constructed to suit your decorative preferences. It is a cute addition to any space despite being a functional area for growing veggies and all types of ornamental plants, with a shocking degree of effectiveness.

Raised beds are supplied with quality soil of preference while keeping in mind all the determining factors like your choice of plants. Here, you are in complete control of all the conditions required for individual varieties.

Designs for Raised Bed Gardening

Raised garden beds can be constructed in a variety of shapes, sizes, looks, and designs. You can choose the portable kind or the fixed kind that can be permanently constructed in your garden or preferred space. They can be constructed with plastic, galvanized iron sheets, wood, or recycled brick stone. It is also possible to build a DIY raised bed with scrap materials readily available at home and at your desired height to achieve both design novelty and convenience.

If you are usually too busy or don't have the required space to pursue your gardening interests, raised bed gardening is the best alternative for you. It can be constructed to fit into your available space and is productive enough to give you contentment and pride in your gardening abilities. A bountiful harvest is expected even at the first attempt. If you're a kitchen gardener, a raised bed will be your secret weapon.

Benefits of Raised Bed Gardening

1. Instantly improved soil: Do you battle with clay-ridden soil or other undesirable soil conditions? Instead of wasting many farming seasons trying to fix your soil, you can simply create an almost perfect cultivating environment instantly with this gardening method. Placing your raised bed on the ground and filling it with high-quality soil is the solution you've been looking for. With

accessible and loose soil, maintaining good cultivation conditions will be a breeze.

2. More productivity: Many gardeners have attested to the ability of raised beds to produce twice as much yield as ground beds; plants thrive in rich, loose soil that allows their roots to penetrate with ease. There is also the benefit of good drainage and aeration. Raised beds prevent soil compression and ensure that the nutrient-rich soil amendments are in place and concentrated for crop nourishment. This promotes dense planting, which leads to more plants in less space than in-ground beds.

3. Longer cultivating season: Soil above ground stays better drained and warmer, thereby extending the growing season.

4. Efficient space: Most raised beds range from three to four feet wide in size, which makes them a great choice for the urban gardener and perfect for small spaces, which allows you to reach all your plants without having to stretch too far or step in. It also enables you to take full advantage of all the planting space.

5. Provision of plant protection: Your plants stay safely away from the annoying threat of human feet and pets.

6. Provision of pest barrier: Raised beds provide a sturdy defense against pests with a plant-based diet, such as snails and slugs. Tall sides guard against non-burrowing critters, and blockades like a hardware cloth can be kept underneath to prevent burrowing root-eating pests from getting to your plants.

7. Less emergence of weeds: Gardening with raised beds results in densely planted crops, which leaves little room for the growth of weeds. Many gardeners fortify this anti-weed fort by placing a weed barrier fabric underneath the bed. If weeds make it into your garden, they will be easy to remove because of the looseness of the soil.

8. Ease of accessibility: Ground beds are associated with a lot of bending, which has been eliminated by raised beds. Bye-bye backaches. The sides can be constructed to allow you to sit down while harvesting or tending to your crops.

9. Aesthetics: Raised beds also serve as an added architectural component because they can be built to be visually appealing. They can also create boundaries, focal points, and symmetry.

The Ideal Size for a Raised Bed

Several factors dictate the size of a raised bed, such as space limitations, physical convenience, and soil conditions; however, it all comes down to two things. Let's take a look at them.

1. Width and length: Mapping out the dimensions of the frame, first consider the constraints of the garden's space, which also includes room to walk around the raised bed. The second thing to consider is accessibility. You should be able to reach the center of your garden from both sides without the need to compress the soil. For some people, this means restricting the width to no more than 4 feet. If accessibility is limited to one side, restrict the width to about 3 feet. The length of your raised bed can be limited only by the lack of building materials and garden space.

2. Height: Most raised beds can be as high as 6 inches to 12 inches, or even 36 inches. The height depends mostly on how bad the underlying soil is. The worse it is, the deeper the bed should be to increase the amount of good soil provided to plants. Also, the deeper the soil, the better the root development. Deeper beds can contain more soil, which automatically means more moisture, which will reduce the time needed to water the plants.

The Varieties of Raised Bed Gardens

There are different kinds of raised beds in existence, including raised bed kits that can be purchased in the market. These kits can be constructed completely out of recycled plastic, cedar, composite wood, and galvanized steel. There are also raised bed designs that are elevated, which will save you the stress of bending every time you need to tend to your plants.

There are raised bed options that last longer. They can be constructed in mere minutes and last for a long time. These kinds are made of composite or recycled wood and usually sport a fancy cedar appearance. Kits like these will not splinter or rot, but if the goal is permanence, the raised bed should be constructed with stone or concrete.

Raised garden beds can be constructed to different heights, the least usually being 6 inches. While deciding on the perfect height for yourself, remember that more height equals more depth, and that means more soil, which will give the roots more room to grow. Usually, the roots of garden crops can grow to about 12 inches, so remember that. Deeper beds also retain more moisture for a longer time than shallow beds, which means you won't have to water the plants as often as in-ground beds.

If you have any issues bending over, it is suggested to use waist-high garden beds. Just remember that these beds consume a lot of compost and topsoil, and it can be a little expensive, but it will give you the chance to continue gardening without the concern of needing medication afterwards.

Ready-Made Raised Beds

A ready-made raised bed is one of the easiest things to assemble. You need only to unpack it, pour in soil, and start cultivating. There are different designs for raised beds in this category so ensure that you select ones that come in a convenient height, are big or small enough for your available space, and complement the existing style of your space.

• Ready-made metal raised beds: If you would like to give your garden a retro look, these pre-made metal beds fit the description. They are usually created from aluzinc steel panels, with stabilized safety edging and stainless-steel fasteners. They can be found in a variety of shapes and sizes, and are great for planting cut flowers and annual vegetables. They complement a balcony outside the kitchen door or even the patio.

• Wooden raised tables: These raised beds are perfect if you want to increase the height of your crops but don't want to fill the entire raised bed with soil. They work great for short-term annual crops that possess shallow root runs like salad crops. They have also proven productive with strawberries, reducing the risk of pests like snails and slugs. They are a nice addition to patios, balconies, and small gardens. They also come with an ease of maintenance and the most comfortable height to work at without straining your back. You'll mostly find them in dimensions like 4 by 4 feet or 2 by 4 feet, and 1 or 2 feet in depth. Most tables are constructed with a thick non-woven polypropylene fabric liner to help with water retention and ensure the durability of the timber.

• Raised trough beds: These are almost like raised wooden tables in that they both provide a comfortable height. They are also called mangers and come in a variety of sizes. The one advantage over wooden tables is the extra depth provided by raised beds. This means you can cultivate deeper rooted plants like parsnips and carrots. The outer, shallower edges can be used to plant shallow rooting vegetables. A liner should be placed on the sides and base of the inside to help retain water and shield the wood from damage.

The majority of pre-made troughs come with a three-year guarantee. A trough is great for small spaces like balconies or courtyards. Placing it close to the kitchen enables easy and quick harvesting of herbs and salad crops.

DIY Kits

Everyone enjoys a good DIY once in a while, especially ones as simple as self-assembly raised beds. They are built in a variety of sizes and shapes to suit different budgets so let's consider the options.

Wooden board kits: This is the most popular type of raised bed kits. They are a great value for your cash and are built in many widths and lengths to give a reasonable degree of garden flexibility. Wooden boards are generally 6 inches in depth, but can also be obtained in 8 inches. They are about five tiers tall, so five 6-inch-deep wooden boards guarantee a 2.5-foot-tall, raised bed. This is a comfortable height for gardeners in wheelchairs.

The wood used usually comes in different types according to quality, but it is commonly built with Scandinavian softwood treated with a non-toxic preservative. It comes with a thickness that ranges from ¾ to 1.5 inches to ensure a sturdy structure that withstands the weight of the soil. A wooden board is great because it is fitted with hoops and covers to protect the plants from pests and frost.

Wooden corner beds: These are the smaller version of the wooden raised beds, which makes them perfect for people with small spaces. They can fit into the tiniest of spaces, thereby increasing the number of vegetables that can be cultivated in the back of the garden. Like the wooden boards, they are available in a variety of sizes and heights.

Hugelkultur: This is a northern European horticultural concept, and its popularity is on the rise as more people are searching for self-sustaining garden methods, lessening the need for constant feeding and watering of the plants. It is a system that depends on the decomposition of wood. This kind of raised bed garden works like a sponge, withholding and releasing nutrients and moisture when necessary.

Hugelkultur involves placing rotting logs, sticks, and branches on top of each other to form piles, and layering soil over them so they resemble little hills. The tops and sides are then used to plant crops. The hills can be built to be large by burying entire tree trunks and leaving them to decompose, or small by burying only a bundle of sticks.

How to Build

1. Remove the turf from your preferred location. Preserve the sod.

2. Dig the underlying soil until it is 12 inches deep. Preserve the soil, separating the subsoil from the topsoil.

3. Place rotting and unrotting timber and logs in the pit to form a pile, with the largest forming the base.

4. Stack the wood until you reach your preferred height.

5. Place the turf over the pile of logs and top with subsoil. Make sure to get some soil between the gaps in the wood.

6. Layer the top and sides with compost or topsoil. Use a rake or your hands to sculpt the soil to form a mound shape. Place logs at the edges to keep surrounding weeds from creeping into your garden bed.

Keyhole Raised Garden Bed

The keyhole garden bed has its origins in Africa, but it is now a worldwide horticultural concept. It is a circular raised bed with a notch cut out for accessibility and maintenance. From a bird's eye view, it is shaped like a keyhole, hence its name. Right in the middle is a compost bin that can be reached through the notch. This compost bin works to supply the surrounding soil with moisture and nutrients.

Due to the compact size of the structure, wires can create frameworks over the bed, which are used to train runner beans, sweet peas, and other climbing plants. They also function as support for shade cloths or netting during hot weather.

The outside walls were traditionally made with stones and rocks in Africa due to their abundance and ability to absorb warmth in the daytime. However, bricks, lumber, empty paint pots filled with soil, and corrugated metal are being used today as building blocks for the garden wall.

How to Build

1. Clear your preferred location, including space for movement around the bed. Remove all perennial weeds in sight.

2. Put a bamboo cane with a string tied to it at the midpoint of where the raised bed will be constructed, tie the other end of the string to another bamboo cane with 5 feet between them. This guide will outline the 10-foot circular raised bed.

3. Make a notch in the circle. It should be roughly ⅛ of the total space.

4. Break up any compression by using a fork to dig over the soil. Then build the outer wall. The traditional height is approximately 3 feet, but it can be adjusted to suit personal preferences.

5. Now build a compost bin at the center of the garden. This is traditionally constructed by weaving flexible canes or sticks together. Bamboo and willow are perfect choices. However, it is much easier to form a tube with a wire mesh or chicken wire, creating a diameter of 2 feet and a height of 4 feet. Keep it secured by pushing bamboo canes through the mesh.

6. Use straw or cardboard to line the insides of the external wall, then layer it with moist biodegradable material.

7. Place alternative layers of green and brown waste board materials like kitchen scraps and cardboard to the compost bin. They will ensure that your plants get the necessary nutrients and

moisture. Make sure not to fill it to the brim to leave space for new materials.

8. Your raised bed garden is now ready for use. Try not to water the plants too regularly. This will force the roots deeper into the center of the bed, making them self-reliant.

Woven Raised Bed Gardens

This raised bed design is great for cottage gardens and is a beautiful feature all by itself. Using supple branches to build structures is one of the oldest methods of construction.

How to Build

1. Clear your preferred location of unwanted materials, and mark the shape of the bed on the ground using string, flour, or sand.

2. Drill wooden stakes into the ground using a sledgehammer to signify where the corners will be. Hazel stakes should also be placed every 20 inches along the sides. Oak, chestnut, and willow may be used. Char them over the fire briefly to make them harder and more durable.

3. Now weave some young willow branches between the posts; ensure it is tightly woven. Stop when the raised bed is high enough.

4. Line the inside of the wall with a black plastic sheet or horticultural fabric liner to extend the life of the branches by a few years.

Recycled Pallet Raised Beds

Pallets are great for recycling because the wood is durable and strong. They look a little rough, but they can be filed and painted to look funky or chic. If you are a DIY enthusiast, you can use the leftover pallets to make complementary benches, tables, and chairs. Note that this raised bed must be constructed.

How to Build

1. You will use four pallets of the same size to form the sides of the bed. If you need to reduce the height of the pallets, don't forget to wear gloves when sawing.

2. Two or three more pallets will be required. Remove the slats from there using a crowbar.

3. Place the first four pallets on the ground and screw the slats over the gaps.

4. Place the four pallets upright and hold them together by screwing metal corner brackets. You should have a box shape when you're done.

5. To protect the sides from rot, line the insides with a plastic sheet or landscape fabric.

6. Give the edges an attractive finish by attaching extra slats to the top edges.

7. File the outside of the box and paint it whatever color you'd like, using an exterior undercoat before exterior wood paint.

8. When the paint has dried, pour in your soil and compost, and it's ready for use.

Recycled Materials

One person's treasure can be another person's trash. You can use almost anything to construct a raised bed, and most of these materials are very affordable or free, environment-friendly, and very fashionable. With recycled materials, you can create looks that range from shabby chic to bohemian, making it seem like you hired expensive garden designers to give your space a makeover. The materials you can use include:

Trash cans: Metal trash cans and old plastic cans are a great option for raised bed construction. About four drainage holes should be made in the bottom and the cans filled with compost to make an ideal cultivating bed.

Old bathtubs: Know any local builders? Contact them for old metal or tin bathtubs from houses that underwent renovation. Simply fill them with soil and get planting.

Old wheelbarrows: No need to trash an old wheelbarrow when you can transform it into the most productive garden. It makes a stylish mini-bed which can be moved around easily. You have only to drill a few holes at the bottom for drainage and fill the barrow with high-quality compost.

Old boats and scrap cars: Raised garden beds can even be constructed from old scrap cars, as shocking as it may sound. Simply pour enough compost into the hoods or trunks, or inside the bodies if you are dealing with a convertible. If the windows and windshields are okay, the inside of the car should make a great greenhouse.

Builder bags: Large builder bags, also called bulk bags, are a fantastic material for the construction of raised beds for many reasons. They are very mobile, easily obtainable, free, incredibly porous, and are an aesthetic addition to the garden. Bulk bags are the large white bags used by builders to transport materials like gravel, compost, and topsoil to building sites and domestic houses. You can use them as they are or construct a wooden frame around them to give them more support and beauty. Simply place them at the desired location and fill with soil.

Rooftop Gardens

These must be ultimately raised beds. They are the perfect solution for the urban gardener without a lot of garden space and a colorful and productive addition to any building. They also attract butterflies and pollinating bees and are a great makeover for unused space.

If you intend to build a rooftop garden you can walk into, consider how heavy the garden is relative to the roof structure. A structural engineer should do this math for accuracy. You will also review building regulations if you'd like to convert a window to a door for access to the garden. If you intend to make any structural changes to the roof, and of course with the actual gardening, you may need to get planning permission as it may affect the privacy of the surrounding neighbors.

Suppose your building doesn't have a flat roof for roof gardening, no need to throw away your gardening interests. There are outdoor staircases in apartments that are laden with containers filled with plants, great as long as they do not obstruct the fire escape in times of emergency. Window boxes are also excellent for raised bed gardening in the absence of a flat roof. The kitchen window is a great suggestion for your window boxes, especially if you intend to cultivate edible crops.

How to Build

Simply select aesthetic containers that complement the current style of the building. A traditional terracotta or aluminum is ideal for a modern house.

Green Roof Garden Beds

Green roofs have become extremely popular as more people are interested in beautifying urban spaces and towns. They help to fight poor air quality and pollution and, like roof gardens, they are a great way to utilize unused space. They may also attract relatively harmless wildlife, which is a huge bonus for animal lovers.

How to Build

1. The first thing to do is get a structural engineer to determine if the building can withstand the weight of the garden.

2. Take measurements of your roof and cut a marine sheet to its exact size.

3. Line the marine plywood with black sheeting or butyl liner and place it on the roof.

4. 3-inch cleats should now be attached to the outer edges of the plywood to form a shallow cultivating frame.

5. Now pour in a mix of general-purpose soil, rock wool, and perlite. This will make the compost substrate lighter than usual to lessen the weight on the roof. The depth depends on the plant you want to cultivate.

6. Make holes for drainage at the bottom edge of the baton to ensure that the bed doesn't become waterlogged. Ensure that you plug the drains with rock wool in the rainy season to prevent your cultivating medium from being washed out.

Chapter Two: Pros and Cons of Raised Bed Gardening

Everything, no matter how seemingly perfect, has pros and cons—including raised bed gardening. It is time to consider the advantages and disadvantages of this method of gardening. We will begin with the notable pros of using raised beds, which includes the advantages they have over ground beds. Several perks are undeniable; however, a few others are slightly more relative, like style and aesthetics. Regardless, raised beds are a popular and great way to grow food comfortably at home. Let's consider why.

The Pros of Raised Bed Gardening

You possess control over the quality of your soil: Raised bed gardening gives you total control over the quality, texture, and condition of your soil. Instead of simply settling for what you have, it is possible to fill your raised bed with a high-quality soil that your plants will thank you for. Soil composition and quality are considered one of the most vital factors in successful gardening.

The best soil for gardening is rich with organic substances, will readily retain water, but also drains well and has a loose texture that easily allows for root growth. This describes the best soil type for farming, sandy loam. Healthy soil also contains some important microorganisms which may not be present in the soil you already have.

Many urban gardeners find their soil undesirable or unsuitable for cultivating food for many reasons. For instance, the soil might have a crummy composition or poor drainage and will need a lot of time, attention, and work for amendment before planting. Also, the soil may have been formerly treated with pesticides or herbicides, which led to contamination.

Some native soils may also be extremely silty. Silty soil lacks air pockets to encourage microbial life, water retention, and structure unless enough effort is put in to saturate it thoroughly. Raised bed gardening gives you the chance to create your perfect growing environment for planting with no need to wait for a long time or put in an exhausting amount of effort.

Determining depth: Raised beds tend to be deep enough for plants that need ample space for their roots. Deeper and bigger root systems mean more yummy plants. This benefit, however, will vary according to how deep you'd like to construct your beds and the structures you install them on, if any.

Generally, it is recommended to stick to a minimum depth of one-foot-tall, raised beds. Raised beds are blocked by a solid bottom or weed barrier fabric. It is advised to construct beds with a minimum depth of 18 inches. Certain plants will thrive in not-so-deep soil, but the most common garden plants–like peppers, eggplants, kale, and tomatoes–will require much deeper soil. Deeper raised beds also have great moisture retention and are better shielded from flooding than ground beds.

This does not mean that ground beds are incapable of having deep soil. However, the composition of the soil might be an issue regardless of depth. Great soil depth without equally great soil composition won't create an ideal growing condition for your plants.

Raised beds are more convenient: Many gardeners usually prefer raised beds because of how comfortable they are compared to ground gardens, especially for their knees and backs. They are easily accessible for gardeners who use a walker or wheelchair or simply have issues stooping or bending. You might need to get on your knees sometimes with some raised beds, but that can be done with a padded kneeler, and you will be less hunched over. As I mentioned earlier, raised beds can be constructed to your desired height, so bending or kneeling can become a thing of the past.

Also, if your mobility is limited or you suffer back issues, it is recommended to construct your raised beds no wider than three or four feet and as long as you'd like. A wider bed will require more bending and leaning to reach the center. There are also raised beds mounted on legs for even more comfort.

Raised beds are mostly pest-proof: Cultivating edible or ornamental plants in raised garden beds ensures an added protective layer against pests. The height of the bed and frame both serve as effective barriers and potential discouragement for plant-eating pests like rabbits, slugs, and snails unless they're determined to eat your garden. Regardless, it isn't difficult to add floating row covers and hoops to block them completely. This has also proven effective against birds, neighborhood cats, squirrels, skunks, and much more.

Netted row covers and hoops can keep pests away from ground beds, but not the kinds that burrow through the soil. For this reason, raised beds are considered a lifesaver by many gardeners. Gophers are a real issue because it is almost impossible to cultivate food in ground beds without gophers killing or eating them. Depending on how they are constructed, raised beds can stop destructive pests, like moles, voles and gophers from destroying your plants.

The bottom of your raised bed can be lined with galvanized hardware cloth to protect your crops. You can also make DIY hardware cloth gopher baskets if you are planning to plant fruit trees. This method protects your plants without requiring you to engage in a never-ending battle by endlessly purchasing traps and poisons to deal with pests.

It is also possible to construct your raised bed high enough to discourage chickens or dogs. Chicken wire is more affordable and mostly used rather than hardware cloth to make gopher baskets or line the base of raised beds. I don't recommend chicken wire, because it tends to disintegrate over time, and it can be chewed through by certain pests. Hardware cloth is a better option for raised beds in areas where burrowing pests are a hazard.

Less weed growth: Raised beds also have the benefit of reducing weed intrusion, unlike ground beds. First, raised beds that have been filled with fresh weed-free soil are very unlikely to grow any weeds, unlike ground beds and native soil that may contain weed seeds and weeds themselves.

The height of the borders in raised beds keeps weeds from creeping into your garden from the surrounding pathways. It is also possible to install a weed barrier at the bottom to keep invasive weeds from creeping into your beds. This must be done before adding soil, or the purpose will be defeated. Weed barriers include cardboard, weed barrier fabric, and so on.

If the preferred location for your raised bed is slightly weedy, before installation, line the bottom with unwaxed cardboard to help kill most or all weeds. Despite the effectiveness of cardboard, some situations require methods that are more effective and durable. Commercial-duty landscape fabric will keep weeds, especially crabgrass, from getting comfortable in your raised beds.

Raised beds are really beautiful. This does not mean that ground beds are not aesthetically appealing; however, the extra visual interest created by raised beds is simply gold. They can create dimension and a well-structured cultivating area. Planter containers of various heights, shapes, and sizes can be organized to create attractive and unique garden designs. Wood planter boxes are also beautiful even when they aren't growing anything, especially if they are wrapped in solar string lights.

It is relatively easy to maintain the aesthetics of a raised bed garden, unlike ground beds. Their borders and edges keep pathway ground cover like bark mulch or gravel from spilling into the garden itself.

They can be placed anywhere: Raised beds can be placed in a variety of locations. Just like containers and other pits, raised beds are very adaptable, with a few being mobile. Ground beds are fixed, so they are limited to only their current location, and that may or may not receive proper sunlight or be level.

Raised beds can be added to a balcony, patio area, terraced into the side of a slope or hill, or even constructed on a rooftop. Technically, they can be created anywhere with good exposure and sound structure. For example, you can construct a few raised beds in your driveway during spring because it receives the greatest amount of sun in the afternoon.

If you plan to set up a raised bed on top of a solid surface, like a balcony or patio, you need to consider adequate drainage, critical for every good, raised bed. The bed also should be constructed with some kind of bottom to keep the soil in place. If not, the soil will seep out slowly, creating a mess you don't want to deal with. One way to do this is to line the open base with geotextile fabric, or you can simply choose a fabric raised bed or a bulk bag.

The Cons of Raised Bed Gardening

As can be seen, there are many respectable pros of raised bed gardening. However, certain potential drawbacks are worth placing under consideration. Let's look at some of the notable ones.

Raised beds require upfront cost and more materials: If only stylish raised garden beds appeared out of thin air. Unfortunately, that is not the case. Tools, lumber, a huge amount of high-quality soil, and screws are required to bring this method of gardening to life. The cost of healthy soil and necessary materials can pile up, especially if you are constructing and filling multiple beds simultaneously. Ground garden beds are more affordable and simpler, and although you might still need to purchase certain amendments and compost before getting started, it is nowhere near as much as that needed for raised beds.

One of the few ways to ensure that filling your raised bed is more economical than usual is to purchase compost and quality soil in bulk. You can also build your gardens and install raised beds in stages, dealing with mini-projects over a while to spread out the cost. Another effective way to do this is to embrace the increasingly popular method of hugelkultur. All you will need is space, branches, bark, and/or logs from anywhere around your property.

Raised beds require a good amount of handy basic skills: If you plan to build your raised bed from scratch, you will need to have certain tools, skills, and physical strength. You will also be required to know basic calculations to purchase the correct sizes and amount of materials and design the garden. If you do not own a saw, lumber departments may cut the boards to your preferred length for you. If it's your first raised bed and you find you don't own a power drill, you might need to resort to nailing it together by hand, though I do not recommend it.

Setting up a ground bed garden also requires a bit of muscle as well, but it is not as tasking and more straightforward, requiring next to zero amount of tools. Fortunately, putting together the parts of a rectangular box is one of the most straightforward and simple DIY projects that can be done by almost anyone, so don't be spooked. To simplify things there are step-by-step tutorial videos all over the internet explaining how to set up your raised bed kit. If you don't feel up to constructing your own from scratch or purchasing a kit, there are pre-made raised beds on the market for convenience. They are available in many sizes, heights, and depths.

They are not eternal: Unfortunately, your raised bed will need to be repaired or be entirely replaced, unlike ground garden beds. When it is time for repairs, a lot of work will be required to replace wooden boards, move the soil, or simply change the entire bed and its contents. Fortunately, the lifespan of a raised bed depends on the material it is built with. For instance, raised beds made of bricks or stones will be more durable than beds made of wood or bulk bags. Also, properly-made wooden boxes will last far longer than poorly made ones.

It is always advisable to build raised beds from good quality wood that is at least 2 inches thick like heart redwood or cedar, which are not only termite- and rot-resistant but can last for more than a decade. Heart redwood is easily obtainable on the west coast and is just as cheap as cedar, which is readily available on the east coast. Redwood and cedar are more expensive than Douglas fir, fence boards, cheap pinewood, repurposed scrap wood, thin 1-inch boards, and plywood. However, they are prone to bow and rot. Also, stay away from pressure-treated lumber because they are packed with toxins, especially if you intend to grow edible plants. The cost of high-quality wood is worth the investment when planning a raised bed garden.

Raised beds are not temporary: When you construct or install your raised garden bed, it is a little difficult to alter the layout of your garden space or change the location of beds. This doesn't mean it is impossible to do; it is simply relatively difficult. You will need to dig out the soil to relocate or redesign them, and that can be a lot of work, especially if convenience was one of the major reasons you started raised bed gardening. Bagging up soil and moving wood from one location to another can be a very tiring experience. Compared to raised beds, ground beds can be easily modified without as much effort. All you simply must do is dig up a new space. You can even plow it over and reseed the area if you like.

You may have limited curves and shapes: Maybe you prefer to enjoy the feel of a more flowing, soft, and natural garden. Ground bed gardens leave room for more flexibility with design and creative shapes, forming fewer hard lines than raised garden beds.

Planter boxes are often limited to rectangular or square shapes unless you own the right tools and are handy. However, it is possible to add some softness and flow to your raised bed garden space in a few ways. For instance, if you have cobblestone-bordered cultivating areas, you can plant flowers, billowing shrubs, and construct curved pathways to give balance to the structure.

So, there you have it! The benefits and potential setbacks of raised bed gardening. As you must have noticed, the potential pros of this method of gardening depend largely on your native soil, aesthetic preferences, unique garden space, the prevalence of pests, and the budget. For many gardeners, the benefits of cultivating with a raised garden bed greatly outweigh the cons, despite both gardening styles being wonderful and worthy.

Chapter Three: Selecting Materials and Styles for Raised Beds

Raised bed gardening is becoming increasingly popular, as mentioned earlier, and for good reasons. People are quickly turning back to nature as a source of safe, reliable, cheap, and healthy food. Its popularity has led to innovations, and these innovations have led countless blogs to feature countless construction materials for raised bed gardening. However, not every featured material is suitable for this method of gardening, with some being very harmful to you and your soil if you are not informed.

Materials to Stay Away From in Raised Bed Gardening

Recycling is generally eco-friendly, and in some cases, is an ideal choice for raised bed construction. However, certain recycled materials must be avoided when trying to construct your raised beds.

Railroad ties: This material has been used to build staircases, garden beds, and other landscape constructs all over the United States. Despite its popularity and availability, it does not look worth the cost, especially if you peer deeper into how the wood was chemically treated before it was ready for use.

The most important aspect of these chemical treatments is creosote and its uses. Creosote has been confirmed to be made of over 300 different chemicals, a lot of which are potentially dangerous to humans and can contaminate the surrounding soil. The EPA has issued several warning announcements against the use of railroad ties in any kind of landscape construction, so you should stay away from it, even if you like the way it looks.

Tires: Tires are usually used for growing potatoes or as a creative way to spice up the look of a garden. This has been beneficial by keeping tires out of the landfill. However, there are heavy metals in these tires. These metals can leach into the surrounding soil, contaminating any food being cultivated on them. There have been arguments about the rubber in tires acting as a binding agent, keeping the metals from separating from the tires and contaminating the soil. At any rate, if you are intent on using tires in your garden, for your own safety ensure that you only plant inedible flowers.

Pallets: Pallets are a great material to use in constructing raised beds as long as you are aware of their source. Pallets were originally meant for shipping materials and had the remainders of whatever they transported. Some pallets have also been treated with methyl bromide, an infamous disruptive chemical that can negatively affect endocrine health. A lot of pallet manufacturers quit the use of the chemical in 2005; however, there are still many old pallets in circulation. If you must use a pallet in your garden, always search for a stamp that says "heat-treated" or "HT." If you don't find a stamp or cannot verify if it has been heat-treated, don't use it.

Treated lumber: A lot of gardeners, including the experienced ones, rely on treated lumber when in need of materials for raised bed construction because of its extra protection against rot, bug damage, and moisture. True, treated lumber is more durable than other materials for the same purpose, but it can also release toxins into the soil, contaminating your food.

Over the years, pressure-treated lumber has been developed with chromate copper arsenate, which eventually leaches arsenic into the soil. Most lumber producers today have stopped the use of CCA during processing. Instead, alkaline copper quat and copper azole are being used, and despite not being as toxic, copper can find its way into your soil, which won't be organic.

If your raised bed has been constructed with pressure-treated wood, make sure to give your plants enough phosphorus via your compost. Plants have a higher chance of absorbing arsenic if they live in phosphorus-deficient soil.

The Best Material for Raised Bed Construction

Heart redwood or cedar: Redwood and cedar stylishly enhance the look of your garden while guaranteeing you a natural resistance to bugs, rot, and moisture. These materials will break down with time, but you can enjoy five or more years from a well-built redwood or cedar bed, with a few even surviving for more than a decade.

Cedar lumber is usually used by experienced gardeners in constructing raised beds and for good reasons. It is naturally resistant to insects and rot. Juniperus virginiana, also called Eastern red cedar, is very resistant to rot and is extremely durable even in the soil. The only downside is that the wood can be difficult to work with because of its density. Eastern red cedar is difficult to find, especially in bulk, because it isn't manufactured locally. It can also be very costly.

Thuja plicata, also known as west coast cedar, is not as difficult to work with, although it tends to split when wood screws are used without pre-drilling. It is easily obtainable compared to other kinds of cedar. There are a few concerns about the sustainability of its production practices. Also, a lot of fuel is required to transport it because of the location of its production. It is about five times more expensive than Southern yellow pine but is usually well worth the investment.

Cypress: This type of wood is native to the southeast and easily obtainable in Georgia compared to cedar, although it usually cannot be purchased at discount lumber stores. It is resistant to insects and rot, especially when in contact with soil. It is more durable than regular pine. It might be a little difficult and expensive to order from a lumber store, but if it is grown and milled in your region, it is much cheaper and a preferred alternative to cedar.

Pine: This has its origins in the southeast, and it is the most readily available lumber in Georgia. Southern yellow pine is one of the easiest and strongest woods to use in construction. It is also very affordable and can be obtained in a variety of grades, with the highest grade being the least common and obviously the best. Regardless of the grade, pine has little to no resistance to insects and rot. Its lifespan is shortened when used in close contact with soil. The only pines exempted from this are the ones from very old buildings. Forty-year-old pine wood is amazingly sturdy, dense, and straight compared to modern-day pine. When seeking pine for the construction of your raised bed, reclaim wood from older buildings and barns because they are a great alternative and more organic compared to other building materials.

Hardwoods like oak: Hardwoods are a little hard to obtain in large sizes or quantities, and according to research, they are only slightly better at resisting insects and rot than pine. The cost of some hardwoods is a prohibiting factor. They are also generally difficult to work with after they have dried up.

Organic Wood Preservatives

Commercial wood preservatives have been put under intense scrutiny in the past decade, especially creosote and the other copper-based pressure-treated lumber, like the green stained lumber used in constructing decks. As I mentioned earlier, recycled wooden utility poles and railroad ties should be avoided when constructing raised beds that will be used to grow edible food due to their treatment with creosote. Similar concerns have been brought forward about pressure-treated lumber, although certain modern formulations seem safe for the production of food.

Regardless, USDA Organic Certification guidelines forbid the use of any pressure-treated lumber to be used in direct contact with edible plants. That means we are stuck with relatively few options for treating lumber for raised garden beds. The two most popular products for this purpose are tung oil and linseed oil. They are not only organic but have also been proven to extend the life of wood even in direct contact with soil.

Linseed oil: This is a flaxseed extract that can shield natural wooden products from rot. It is vital to understand the difference between boiled linseed oil and raw linseed oil. Boiled linseed oil is a combination of raw linseed oil and artificial solvents that may not be safe to use in organic systems. Raw linseed oil is a very affordable natural wood preservative. It is not as effective as copper-based preservatives or creosote, but it is completely organic. Note that linseed oil is a source of food for mildew, so don't be surprised when you see mildew growth on wood preserved with linseed oil.

Tung oil: This is a tung tree extract that has proven effective in the preservation of wood. It is more expensive than linseed oil and is usually mixed with toxic solvents that help in application and absorption.

Unpreserved Wood

This kind of lumber makes some of the most beautiful, homey looking pieces, including raised beds. The one thing to remember, however, is that products made from untreated wood are not as durable as the other options. Raised beds made with untreated wood can work for about three years before they need replacing. This has proven itself to be a good economical option for gardeners looking for cheap, durable, and temporary raised beds before making some permanent raised bed additions to the garden.

Rocks

If rocks can be easily obtained from around your property, make good use of them by building a natural raised bed. You might exhaust yourself trying to get the rocks to the preferred location. Still, the initial effort will produce long-term benefits, considering rock raised beds are almost eternal, requiring little maintenance. You will also need mortar to glue the rocks together, at least while building upwards for height.

Remember that this option is only economical if you already have rocks around your property. Buying rocks will only result in more cost and is inadvisable unless you intend to do it more for pure aesthetics than function.

Bricks

This is another stylish option but can be a little pricey depending on the brick you're in the market for, whether recycled or new. Just like rocks, gardens made with bricks last for many decades, requiring little maintenance. A cheaper alternative has gathered a lot of attention recently, thanks to YouTube–cement blocks. However, do not use the cinder block form of cement blocks, particularly the older ones, if they are mixed with fly ash. Contained within fly ash are arsenic, lead, mercury, and so on, which will seep into the surrounding soil and contaminate your food.

Concrete Blocks

These are the cheapest and easiest material to use to construct a raised bed garden. Cement blocks are readily available for free or a little fee. They can be placed on top of each other to make fairly high raised beds. However, if the walls are higher than two levels, use mortar to hold them together to prevent collapse. It is easier to find cement blocks that have been used, but remember that stacking used blocks without mortar is almost impossible.

Pots and Containers

This is a great option for gardeners who live in apartments but are still interested in pursuing their gardening interests. Even gardeners with ample space can reap the benefits from the pots placed in nooks and crannies of abandoned space. This is especially perfect for spreading plants, like certain caneberries and mint, in bigger pots. Don't forget to drill extra holes than those provided by the manufacturer for better drainage.

When purchasing pots for planting, seek ones that are BPA-free to keep BPA from seeping into the soil. This information is usually found at the bottom of the pot.

Composite Wood

Composite wood is becoming increasingly popular for outdoor building projects due to its durability and relative ease of use. It is made of a mixture of pulp or wood fibers and plastic resins and is then shaped in different dimensions of lumber. It is expensive and available in a few basic sizes, especially ones for decking. Typically, it is about three to four times the cost of basic treated pinewood.

The lifespan of composite wood in direct contact with soil hasn't been confirmed. Studies of the effects of composite wood for decking have shown that the wood undergoes many kinds of deterioration similar to traditional wood such as discoloration, mildew, cracking, degradation in light, and mold. Studies have also confirmed that composite wood has more environmental costs than pine or cedar.

Mortared Walls

Mortared walls are a more permanent and more secure construct. Dry stacked walls are not as expensive, easier to set up but also not as permanent. Concrete excavated from sidewalks and driveways is usually obtainable for free. If you get tiny pieces in a uniform thickness, you can use them as great recycled materials for your raised bed walls. There are certain areas in Georgia where granite rubble is easily obtained. If they are angular stones, they can be stacked into walls. Keep in mind that a stone wall cannot be constructed higher than a single level, with mortar to keep the stones together.

Raised Bed Kits

These have gained popularity at local garden centers and with internet suppliers. Often, they make raised bed construction fast and easy. They are typically made of either western cedar or cheap plastic, and while some kits only contain the hardware, others are the entire package. The quality and variety of raised bed kits vary greatly, so be cautious when making purchases, ensuring that you know of the kit you want before a purchase is made.

Regardless of the size of your garden space, raised beds are the perfect organic solution for cultivating healthy and productive crops at any skill level and age. As long as you make use of the right materials for safety, convenience, and style preference, you will be on the path to a more active, healthy, and self-sufficient lifestyle.

Chapter Four: Creating a Layout Plan for Your Space

A functional planting scheme needs a good plan. More time should be allotted to choosing and selecting the proper crops than the main process of cultivating. You must decide on the style you require and then research the plants' requirements.

Raised beds are an efficient way of gardening, but they come with a list of challenges. Proper planning and an awareness of all the factors involved are necessary if you intend to have a functional raised bed garden.

Practical Things to Consider

Lifting certain materials to a certain height is considered hard work by some, even to the point of causing injury or strain if done incorrectly. For instance, a watering can is very convenient when watering plants at a relatively low height, while lifting the same can to a raised bed height can feel uncomfortable. Thankfully, solutions to this exist in a hose or an irrigation system.

It is not easy to lift heavy materials like heavy plants or wheelbarrow loads of compost. However, there are solutions such as scaffold boards or low ramps to help push the compost in little amounts onto wheelbarrows. Or you can always use lighter, smaller containers like buckets to move building materials.

The Perfect Height

There are tall climbing crops like hops and French runner beans, which might grow too tall beyond reach if you cultivate them in upright structures or high raised beds. This will require the use of a ladder to tend to and harvest your crops. Luckily, there are shorter versions of these plants just as productive and easy to grow. Always keep height in consideration.

The Budget

The cost of constructing a raised bed is higher than planting in-ground beds, regardless of the kind of materials you intend to use for construction—whether rocks, lumber, or bricks—unless you have enough of these materials around your property. You also must purchase screws, drills, hammers, jigsaws, and nails to hold them together, and consider the cost of your planting material. If we lived in a perfect world, everyone would have their own bags of homemade compost, but most people will need to import soil and compost into the garden. One good thing is that the initial outlay is always worth the costs because, in time, the beds will become increasingly efficient, producing bountiful harvest every season. At least you won't have to spend money on pest control.

Moisture Retention

The kind of raised bed you finally settle on will determine if you must water the plants more often than you would in-ground beds. This is due to the better drainage provided by raised beds when compared to ground beds, which can be a good thing; however, some drain more than others. Raised beds like the keyhole garden and hugelkultur are designed to retain water instead of losing it.

The Drawing Board

The ideal way to start planning your raised bed design is to draw out the plan on a piece of paper, determining what plants will be placed where. This stage of the planning process saves you from making expensive mistakes, like purchasing more plants than you need from the garden center.

If you intend to plant veggies in your raised beds, then you will need to plot methodically to determine the appropriate position for each vegetable. This ensures that all plants are given the necessary space and prevents the smaller ones from constantly living in the shadow of the taller ones. During this process, you might find you need more than one raised bed so you can engage in crop rotation every year. If your garden is strictly ornamental, you will need to make sure that your preferred plants are suitable for the climate conditions and soil type.

Style: Pick a Color

For an ornamental raised bed, you will also take into consideration the color combinations of the plants. You may find you want a single-color theme in your garden, or you may wish to have colors that project a particular mood like tranquil, calming pastel colors, or vibrant, bright, hot colors. Many gardeners use a color wheel to determine the colors that will go well together.

The Color Wheel

A color wheel is an important tool for determining great color combinations. They are not only used in gardening but also in other areas like fashion, art, marketing, and so on. The rule is that the colors sitting alongside one another, like yellow and green, will be a calm and harmonious combination. In contrast, colors on the opposite sides of each other on the wheel, like purple and yellow, make a clashing combination and a striking impact.

The Importance of Size

When designing a ground garden bed, the placement rules are usually according to size and are fairly easy: The short plants stay in front while the tall ones stay at the back. However, raised beds are a little different because they can be viewed from multiple angles, so consider how you want the bed to be viewed.

If you construct a path that runs around the entirety of your raised bed, it will be seen from all angles, meaning that the shorter plants should be placed at the edges and the taller plants in the center. Remember that the taller plants might throw some shade on the shorter plants according to the movement of the sun, so carefully consider the position of all plants.

If you intend to grow vegetables, consider placing the shade-tolerant plants like lettuce and leafy veggies next to the taller vegetables. Keep sun-loving plants like pumpkins, squashes, and tomatoes close to the edges of the bed, away from the taller vegetables. Remember that plants such as potatoes and carrots require at least half a day of direct sunlight.

The Perfect Depth

Different raised beds require different depths for different reasons. These factors are determined by the plants you intend to grow and whether the bed is situated on a patio, has a board at the bottom, or is in direct contact with the soil below. If the bed sits directly on the soil, the crops with bigger roots will spread beyond the depth of the raised bed into the ground below. Here, the height of the raised bed is not important.

Annuals, Herbs, and Salad Crops

If you intend to grow only a handful of salad crops yearly, you will require a raised bed that is about 4 inches deep. Salad crops tend to have shallow roots and will even be equally productive in a window box, so depth in a raised bed is not a necessity. There are other reasons for wanting to use a higher raised bed besides depth, like aesthetics or ease of maintenance. Begonia, petunia, lobelia, annual rudbeckia, cosmos, Busy Lizzie, and other annual bedding crops require similar conditions and do not possess large root systems, meaning they are fine with shallow soil. Most perennial herbs like mint, thyme, and rosemary have their origins in the Mediterranean region, where they grew in rocky and arid soil conditions, so they will not require the depth.

Deep-Rooted Vegetables, Ornamental Grasses, and Herbaceous Perennials

Herbaceous crops require depth in raised beds because they typically have a larger root system than annual crops. Veggies like potatoes, peas, beans, carrots, cabbages, and so on also require depth. To be specific, a depth of at least 12 inches is required for them to reach their higher point of productivity. They can grow in shallow beds, but you might need to water and feed them more than usual to make up for the lack of depth.

Fruit Bushes and Shrubs

Many shrubs and fruit bushes need a depth of at least 20 inches to grow and develop to their full potential. Like deeper-rooted vegetables, they can grow in shallow soil, but they will be considerably stunted and are likely to have short lifespans compared to the same plants in deeper soil.

Trees

Did you know that when you look at a tree, the part you see above is usually a mirror of what is below in the form of root systems? Now imagine the root system of a huge tree. Remember that trees are typically adaptable, the bonsai tree being a good example of trees curtailing their growth to fit the available space. Certain trees can be purchased on dwarf rootstock, which curtails their overall size. In an ideal world, trees should be granted a depth of at least 3 to 4 feet in a raised bed.

Factors to Consider When Siting Your Raised Bed

One of the major keys to successful gardening is placing the right plants in the right raised bed relative to the position of the bed. If you are dealing with a small garden, you are unlikely to have a lot of options for siting your raised beds, but fortunately, plants for every aspect exist, whether it's a sunbaked, shady, dry, or damp corner bed.

Shady Corner Beds

If you can, and your space allows it, the best site for your raised bed is an open, sunny area. Most plants thrive in maximum sunlight; the more sun exposure to their leaves, the higher their production of sugars, which sweetens any vegetable or fruit they produce. However, if your space doesn't permit you to place your raised bed in direct sunlight for most of the day, no worries because there are plants that love the shade. Leafy plants like spinach, summer salad greens, and cabbages like it cooler. Shady corners mean that the beds are less likely to dry out due to a lot of sun exposure, and the vegetables have less tendency to bolt because they enjoy the cool root system. Many ornamental plants also thrive in shady corners, like ferns, hellebores, Epimedium, and hostas.

Proper Lighting

Before constructing a raised bed, it is advisable to find the area where the sunlight touches the most to maximize the sunlight given to your plants. This might seem obvious but remember that a flower planted directly on a ground bed, especially in small gardens, might be blocked from direct sunlight when elevated onto a raised bed. Walls, roofs, and tree canopies can block a garden bed normally bathed in sunlight when on the ground.

It's easy to understand. You know that the sun rises in the east and sets in the west. When it's midday, the sun is always in the south, and because of this, raised beds facing the south are sunnier and much warmer than the ones placed on the north side of the house. A raised bed is meant to be in direct sunlight for most of the day, so if your backyard is on the north side of the house, but you are fortunate enough to have a lot of space in your front yard, then consider placing your raised bed there. Also, remember that the sun is much higher in summer than it is in winter, so if you plan to extend your planting season, check to see if your garden still gets adequate sunlight even while at its lowest height.

Another way to let more sunlight into your garden is by cutting back overgrown and overhanging vegetation branches. Don't forget to ask your neighbors if they will be okay with reducing the height of a few trees in their garden that affect the amount of sunlight coming into yours. Reducing the height of your boundary fence will also let more light in, even though it might cost you your privacy.

Location

There are also other practical issues to consider when picking a location for your raised bed. If you intend to cultivate herbs or vegetables, then consider placing the raised bed near the back door or kitchen window so you can harvest fresh veggies or herbs easily when cooking. To create more privacy from outsiders, you can consider placing your raised beds on the garden walls to increase the height of your garden boundaries. You can also place them around a seating area or patio to create a sense of privacy. If the plan is to cultivate trees or tall plants, keeping them away from the house is best, so it doesn't restrict your view of the garden.

Provision of Shelter

Just like us, crops often prefer to be protected from the elements. Wind exposure can cause their leaves to decimate. It can also cause the affected crops to dry out quickly as it absorbs most of the moisture in the soil. Strong winds during blossoming periods do not allow pollinating insects to fly around and do their jobs, which will lead to low yields in fruit bushes and trees.

One effective solution is to cultivate tenacious, tough crops capable of tolerating the wind. Plants that thrive in waterfront locations are ideal. However, if your plants are tender, like most vegetables, they will require some protection from strong winds. Many small gardens, especially in town, usually have proper shelter from winds because walls, fences, and hedges surround them. For large gardens, avoid constructing raised beds in conditions that will leave them exposed–atop a hill, for instance.

The perfect windbreak for any garden is a hedge because it lessens the impact of the wind while being semipermeable, allowing adequate air circulation. This is vital because it aids in the prevention of diseases, especially fungus, and the buildup of pests which flourish in still, stagnant conditions. Structures that are non-permeable, like fences and walls, effectively prevent wind damage, but they come with a detrimental catch. Sometimes, the wind may move along the top edge of the fence or wall and take a nosedive right onto the raised bed with more force.

Frost Pockets

So many plants will pay the price if your raised bed is placed in a frost pocket. Frost typically gathers at the lowest parts of the garden because when cold air passes through it, any warm air that arises is replaced. This effect can be worsened if the cold air is not allowed to circulate, and this is usually caused by a permanent solid structure such as a fence or wall at the lowest end of a garden. Seedlings take the brunt of the cold, getting zapped quickly by the frost, while blossoms or young tender shoots will simply wither and die. It also lessens the length of the growing season because the bed will be too cold to grow anything until well into spring.

Keeping your raised bed away from frosty sites will permit early spring cultivation and lengthen the growing season well into autumn. If a frost pocket is unavoidable, prepare to shield the plants and cultivate later in the season to prevent the disappointment that comes with losing your crops to the harsh cold.

Chapter Five: Constructing Your Garden Beds

Simple Steps to Construct a Wooden Raised Bed

1. After picking a location for your raised bed, determine if the native soil is of high quality. If it is, it should be dug out and reserved to fill the bed later.

2. Use a string or rope to outline the perimeter of the bed on the ground.

3. Retaining stakes of a minimum 2 x 2 inch are required for raised beds. These stakes go into the corners of the bed. Place stakes at every 5 feet along the sides for support. Now push them 12 inches deep into the ground.

4. Use galvanized screws to attach the stakes to the retaining wooden boards.

5. Pour the reserved soil or purchased soil into the bed, filling only the lower section. If the grass was removed from the raised bed site, you can place it upside down at the bottom of the bed because it will slowly rot as the season goes by.

6. Now fill the rest of the bed with an equal mix of garden compost and topsoil.

Simple Steps to Construct a Brick Raised Bed

Brick raised beds are more difficult to construct than wooden ones, needing extra practical skills like bricklaying. Don't fret; it's nothing you can't learn. Brick raised beds are unsurprisingly durable once built properly, providing a sturdy, strong bed that is stylish in most locations–backyard, front yard, or garden.

Cutting a Brick

1. Use the edge of a trowel to make a slight groove in the center of the brick.

2. Place a brick bolster in the groove and use a hammer to bang down swiftly to cut the brick clean in half. Don't forget to wear protective eyewear when cutting bricks.

Steps to Construct a Brick Raised Bed

1. Using a string or rope, map out the perimeter of the raised bed.

2. Now build a footing made of concrete to prevent the bed from sinking into the ground. Now make a trench with a depth of 20 inches and width of two bricks. Mix 2 ½ parts sand, 1 part cement, and 3 ½ parts gravel to make the concrete foundation, then use it to line the bottom to a depth of 6 inches. Now, wait for it to dry.

3. Mix 3 parts sand and 1 part cement. Add as much water as needed for an easy-to-use consistency that is flexible enough to coat the entire brickwork, but not too sloppy. The plasticizer helps to keep the cement mixture flexible.

4. The courses of bricks should be laid two bricks wide. Every brick must be laid onto a bed of cement that is only an inch thick. The second layer should be started with half a brick for it to be staggered with the courses below it. This will reinforce its strength and durability.

5. The first three layers should be high enough to reach the ground level. Now, keep stacking bricks until you reach your preferred height.

6. If you have any chamfered bricks lying around, cement them at the top edge of the bed for protection from dampness and for aesthetics.

7. The inside of the walls should be lined with a permeable material.

8. Pour in compost and topsoil.

Steps to Construct a Herringbone Brick Path

1. To make sure the path is level with the surface, dig out the soil at the location of the path to a depth of an extra inch than the bricks you will be using.

2. Now layer the bottom of the path with sand to a thickness of 1 inch.

3. Place the bricks in a herringbone pattern along the length of your path.

4. Bed the bricks into the layer of sand and then fill any gaps with extra sand.

Garden Paths

If you intend to have multiple raised beds, you need to carefully consider the paths between and around them. Paths create the structure and backbone of any garden design and are important in ensuring that the essential elements of the garden are accessible.

Garden paths need to be practical and functional, but should also look beautiful and complement the existing style of your garden and raised bed. For instance, a rustic wood chip path will beautifully complement a formal brick raised bed.

The Perfect Width

If you hope to walk comfortably between your raised beds, the least width for your path should be 16 inches. If you intend to use a wheelbarrow while gardening, the path should be at least 26 inches in width. If you want it wide enough for wheelchair access, the path should be about 3 to 4 feet wide. Remember that there will be no extra room for anyone else besides the one wheelchair. For extra room, make it 5 feet wide.

Types of Garden Paths

Paving slab or brick path: The sturdiest paths are built with paving slabs or bricks which provide a strong base for your wheelbarrow to move along. If you are going for a rustic look, consider laying them into the soil, pushing them down to the depth of the slab or brick so they are level with the ground. You can also lay them on the sand and simply secure them with a brush of dry cement and sand mixture, watering it after to set it. They are available in a variety of sizes and prices to fit most people's budgets.

If you are not concerned about style, check the dumpsters in your area because people are often getting rid of old paving slabs and bricks. Don't forget to check with the builder or owner before going through their dumpster.

Grass path: This is the most affordable kind of path but requires the most maintenance as it will need to be mowed at least once a week in the growing season. If you choose to lay the grass right up to the sides of the raised bed, the edges also require regular trimming. Some raised beds might cast shade over a good portion of the grass paths. If so, shade-tolerant grass seed is the best option to ensure the grass is green all year long.

Woodchip mulch path: This option is relatively cheap. To construct one, place gravel boards on either side of the path and hold them in place with wooden pegs. Gravel boards are usually 6 to 8 feet long, 3 to 4 inches wide, and 6 inches high. They work to retain the woodchip mulch on the path and keep it from spreading onto other flower beds and lawns.

Now, place a layer of ground-suppressing membrane on the path, and hold them down with metal tent pegs. Finally, the path should be coated with a 2-inch layer of woodchip mulch and raked to level. The wood chip mulch will need to be topped off regularly as it rots or washes away.

Gravel path: The construction of this path is relatively straightforward and cheap. One con is that spilled soil or compost is hard to pack and tidy away. To construct, dig out a base about 6 inches deep. Place treated lumber at the edges to prevent gravel from spreading to unwanted areas. Wooden pegs should be driven in at 3-foot intervals to keep the lumber in place.

Use a roller or a Wacker plate to compact the ground or simply tread it flat if it covers only a small area. A layer of hardcore should be placed at the base of the path, then layered with sand and topped off with a layer of gravel 2 inches thick. Level it with a rake, and you're done.

Common Mistakes to Avoid in Raised Bed Gardening

Poor arrangement: Siting your garden in the wrong area is a big mistake and almost impossible to fix when using a raised bed. The box is usually hard to rearrange or move once you have filled it with soil, installed the water system, and planted your crops.

To avoid this, the first thing to consider when picking a location for your raised bed is the sun. If the orientation is east-west rather than north-south, your plants are less likely to receive the required sunlight for their development. Vegetables need at least six hours of sunlight every day.

Planting shade-loving plants in the sun or vice versa is another issue that most beginners face. For instance, tomatoes require six or more hours of direct sunlight every day. Chiles, eggplants, and other herbs will be happiest in the sunniest part of the raised bed, while peas or lettuce will prefer shady placements.

That said, plants placed on the south side will get the most sunlight, but ensure that they are short enough plants so they do not restrict sunlight from the other crops.

Unsuitable construction materials: Most raised beds are constructed with wood, but they can be built with a variety of other materials. Confirm the materials are safe for use near your crops, particularly edible crops. Safety standards and health regulations usually vary depending on the region or state.

Toxic materials like pressure-treated wood or chemical-treated lumber should be avoided when sourcing for materials to construct your raised bed. Creosote or other harmful chemicals might be contained in older materials, so avoid those too.

Search for sustainable and locally sourced options that haven't been treated, are resistant to rot and long-lasting, and FSC-approved (Forest Stewardship Council). This will ensure that your raised bed stays functional and wonderful for a long time.

Picking the appropriate size: Try not to go with a raised bed bigger than you need. Your beds should be just the right size for easy access and convenience. The recommended size for a typical raised bed is not over 4 feet wide, to enable the gardener to reach the crops in the middle.

If you site your bed close to a fence, it will benefit you to reduce the width to below 30 inches. Also, ensure that you leave enough space between raised beds, at least 2 to 3 feet. Every gardener should be able to walk through pathways and between beds comfortably.

Watering: Watering your plants too much is another common mistake as your plants can drown and rot. Watering them too little is also problematic. Suppose you are unsure of the amount of water required by your raised bed; no need for any guesswork because you can invest in an irrigation system with a smart controller. It has moisture sensors that automatically detect and adjust the amount of water in your garden.

Your irrigation system need not be state of the art or expensive to function properly and save you a lot of time. However, if an irrigation system is out of your budget your crops need not suffer. All you must do is closely observe the soil. Once it appears hard, it's watering time. If you cannot tell by looking at it, grab a handful of soil and squeeze it into a loose ball. If it sticks together, the soil is moisturized enough. Certain plants act as moisture indicators. A good example is lettuce, which quickly wilts when dehydrated. Consider planting indicator plants to help you properly check the moisture content at a glance.

If you don't have an irrigation plan in place when constructing your raised bed, you will have to water the old-fashioned way, with a long hose or watering can. Another option in the absence of an irrigation system is to place a rainwater barrel close to your raised bed for convenience's sake.

Poor soil quality: Just like a living organism, the soil goes through changes and evolutions. Its conditions are affected by rainfall, drainage issues, or runoff. Certain plants feed off the soil more intensely than others. It is vital to pay attention to the kind of soil you use for your garden–it's mineral levels, pH, and the necessary organic matter required to give it a boost.

The kind of soil you put into your raised bed is an important aspect of your crops' future happiness. Avoid using normal potting soil for your garden because it drains rapidly. There are available raised bed soils on the market that are more effective.

You can purchase a DIY testing kit from a hardware store in your area to test your soil annually. This kit helps you discover the kind of soil you need, considering the crops you want to grow. Test your soil before cultivating and throughout the lifetime of your raised bed. To obtain the most effective soil, mix it with equal parts of organic compost. Your plants will surely take advantage of this nutritious addition.

Chemicals: Making use of the wrong chemicals directly on or near your raised beds can severely affect the productivity of your crops. You may think that it is safe to use these chemicals in your garden but away from raised beds, right? Wrong. The wind can carry all that toxicity to your beds, harming your plants.

Chemicals containing herbicides can linger in the dirt for many years, poisoning the soil. Sure, it is important to get rid of the weeds and grass, but if you get too close, you will lose your plants. These chemicals become extra dangerous in the rainy season because runoff water can transport them to other areas of your garden.

Moral of the story: Stay away from toxic herbicides. Instead, mix equal parts of vinegar and hot water to get rid of the weeds and grass. Simply spray the offending plants with the mixture once a day every day until the weeds wilt and turn brown, then uproot the rest by hand.

Your garden pathways will probably also grow grass and weeds eventually, but instead of spraying them or mowing them, which are also good ideas, you can simply build a barrier. Flatten as many cardboard boxes as you need and layer a little mulch on top of your barrier. It is an easy and more durable solution than other options.

Lack of preparation: Proper preparation of the beds between growing seasons will ensure healthy and bountiful harvests. If you fail to prepare your soil for the next season, the crops you grow may be diseased, stunted, or not grow at all.

Rather than plant the same crops in the same position every year, you can cultivate healthier crops by practicing crop rotation and avoiding planting crops of the same family in the same spot or near each other year after year. Fungal diseases, soil infertility, and common pests are issues faced by different plants; to prevent a problem from spilling over to other plants, alternate the positions of your crops every year.

Wrong vegetable choices: Picking suitable vegetables but in the wrong combination is an error that can be corrected later, but choosing the wrong varieties? Your planting season might be more difficult than you hoped. Let's assume you start with a tougher vegetable like asparagus. If you're a beginner, you might become discouraged by the long wait of two or three years for it to produce a harvest. Another mistake beginners make is cultivating a cool-weather crop like cabbage in the wrong season.

To avoid this, begin with the vegetables that are simpler to grow while you figure out the crops that work well for your gardens, like basil, bell peppers, tomatoes, and zucchini.

Ensure that the vegetable options you select are not just easy to cultivate but will also be suitable for yourself and your family. It is pointless to grow lettuce if there are any allergies in your home. Select the vegetables that will be consumed the most, and you will be more likely to be interested in the variety held by your garden.

It is also vital that the options you pick thrive in your yard, because some might not. Certain vegetables are more prone to pests, don't thrive in humid environments, or cannot withstand sudden changes in temperature throughout the year. Always consider the weather of your location.

To start your first year cultivating an all-herb garden with herbs easy to grow both outdoors and in, here's a list of the simplest herbs to start with:

1. Thyme
2. Parsley
3. Basil
4. Mint
5. Cilantro
6. Oregano

Not using row/seed indicators: Mark your rows and mark the spots where you plant every new addition to your raised bed to prevent overcrowding. It is easy to lose track of the exact spots you put seeds in, but using an indicator will prevent you from replanting over seeds because seedlings can sometimes be mistaken for weeds.

For convenience's sake, place a label where every plant is. Purchase small plastic tags and stick them in the soil to label the plants.

Gardening is an adventure where you never stop learning. Even the most experienced gardeners make mistakes now and then, but learning from these errors is the only way forward. Plus, there's no harm in trying out something new!

Chapter Six: Choosing Cultivars (Plus Tips for Organic Gardening)

This chapter will guide beginners in the field through the sometimes-frustrating process of seed selection and garden preparation. It is also for experienced gardeners who may be dealing with seed-shopping addiction.

The first thing to do is understand the jargon because, like other trades, seed catalogs have certain phrases and terminologies that are crucial to understand. You may know the meaning of the certified organic label (and it is okay to be unaware. This label ensures that these seeds have been cultivated in organic soil and have been cared for and processed according to the guidelines issued by the USDA). However, the description of seeds goes a little further than that. Here are a few important terminologies you will discover in the catalogs:

1. Cool-season plants: These are plant species tolerant of the frost. They thrive in fall and spring when temperatures in the daytime are in the 70s or 60s, and nighttime temperatures range from 30 to 40 degrees.

2. Warm-season plants: These flourish from late spring well into early fall when the temperature in the daytime is over 80 degrees, and nighttime temperatures are typically over 50 degrees.

3. Days to full growth: This is the average number of days required for transplanted crops started indoors to reach maturity. It is also the number of days required for seeds sown directly in the garden to mature.

4. Disease resistant: These two words are the most crucial to look out for when going through seed catalog listings. Some listings contain acronyms like VFNTSt or VFN, which is a shorthand tag for the particular disease that a variety is immune to. A VFN berry is immune to fusarium wilt, nematodes, and verticillium wilt. Every seed catalog should come with a cipher for its disease-resistant codes.

5. Heirloom: These are traditional varieties that have been passed down through generations, instead of being produced by modern seed farmers. The majority of the heirlooms available today predate the 1940s.

6. Open-pollinated: These varieties have been naturally pollinated by insects or the wind, instead of the controlled pollination methods used by experienced breeders. A seed from an open-pollinated plant can be reserved yearly because it will germinate true from seed, that is, it will look almost identical to its parent plant.

7. F1 hybrid: These seeds are a product of a deliberate crossing of two varieties. F1 hybrids give rise to plants with greater consistency in appearance, size, and other characteristics. The seed produced cannot be reserved and planted the next season again because it will grow into plants that have significantly different traits than the original plant.

8. Non-GMO: This seed is not created using any genetic engineering methods. Examples are heirlooms, F1 hybrids, open-pollinated seeds, and any variety bearing a certified organic label. Most of the retail seed companies sell their products as non-GMO, but honestly, there are no GMO seeds available for sale to home gardeners. Non-GMO seeds are almost strictly used in North America for industrial agriculture.

9. Pelleted variety: These are seeds that have been layered with a biodegradable substance to increase its size and make it easier to cultivate. It helps to lessen the overplanting of small seeds, like carrots and lettuce.

Additionally, be on the lookout for phrases or terminologies that might point out a trait you think will be useful to you. For instance, a "bush type" bean is a kind of bean that grows stocky and low, which makes it a perfect choice for gardens with limited space. There are also "patio tomatoes," which grow stocky and short, and have been bred to flourish in a container or pot. Speak to your neighbors or friends about the varieties they prefer or contact your local cooperative extension service officer for suggestions on what seeds to plant.

The Link Between Climate and Vegetables

It is okay to select some seed varieties that will require more time and attention than others because they aren't designed to thrive in your climate. It is recommended, though, to purchase seed varieties that have no complaints about the climate in your region.

The first thing to focus on is discovering your first gardening window. This is the approximate number of days in a year free of frost. This is important because the growth of many crops comes to a halt when the temperature plummets below 32° F. To do this, mark the day the frost ends in spring and the day of the first frost in fall in your area.

Some vegetables like basil, melon, tomatoes, beans, corn, cucumbers, and so on are warm-seasonal. They are unlikely to survive the first frost of the year, so they must be cultivated and harvested during the gardening window. Other vegetables like cabbages, potatoes, carrots, and so on are cold-seasonal. They are more likely to survive a mild frost. However, they rarely make it through the summer unless they are in the north or coastal regions where the summers are cooler. If you live in an area with hot summers, only cultivate these kinds of crops in fall and spring.

When you purchase a seed packet, check for an indicator that reveals the 'days to full growth' of that seed variety. You are unlikely to find that information in the catalog listings as they only tell you the regions suited to the crops. Crops that require 90 days or more to maturity are unlikely to survive in areas with short, cool summers because they are heat lovers. The number of days to full growth written on the seed packet will simply show you the intensity of warmth or coolness needed by the crop to thrive, not necessarily indicate the exact number of days required by the seed to mature.

Tips for Organic Gardening

Cultivating organic edible plants means you and your family can enjoy tasty, fresh, and healthy harvests free of pesticides or synthetic chemicals. Some of the basic tips for organic gardening are the same as in nonorganic gardening: Place your plants in an area that gets maximum sunlight for about 6 hours a day or more. All gardens need to be regularly watered, so ensure that you have a hose or spigot that will reach all parts of your bed. Now, we will look at the tips particular to only organic gardening.

Begin with mulch and organic garden soil: To enjoy the harvests of a healthy organic garden, you must begin with healthy soil. The most vital part of a soil's makeup is the organic matter such as compost, manure, or peat moss, which is a product of decayed microorganisms of former plant life. This decayed matter supplies the crops with the necessary nutrients for their survival. It is possible to create your own compost by setting aside a bin or area where the organic matter will be left to decay. Or you can just purchase it in bulk if you own a lot of raised beds. You can also use bagged compost, available at home improvement stores or garden centers.

Limit the spread of weeds by layering a 2-inch thick coat of mulch on the soil. It forms a barrier that keeps weed from receiving the sunlight they need to germinate. This layer of mulch also prevents spores of fungal disease from drifting onto plant leaves. Use an organic material like weed-free straw, newspaper, or cocoa hulls as mulch so it will decompose, contributing to the organic matter in the soil.

Don't ever settle for inorganic garden fertilizer: Using fertilizer in your garden will ensure that your crops grow faster and produce larger crops. The types of organic fertilizer you should look for are well-decayed manure from plant-eating critters like horses, chickens, sheep, etc., pre-packaged organic fertilizer purchased at your neighborhood garden store or online. A variety of organic fertilizers can also be found at home improvement stores and garden centers. Note that if your soil is already rich in nutrients, you can consider skipping fertilizer. Too much goodness can attract pests as they will be drawn to the lush growth of your crops.

Tips for seedling shopping: When you shop for seedlings, you should choose plants that have no yellow leaves and a healthy color for the species. Steer clear of wilting or droopy leaves. If you're in the market for transplants, carefully tap the plant out of the container or pot to ensure that the roots are white and well developed. Stay away from plants that already have flowers or are budding. If they can't be avoided, pinch the flowers and buds off before planting to make sure the plant directs its energy into forming new roots.

Rotate your crops every year: A lot of closely related plants usually suffer the same disease, so it is necessary to avoid growing crops in the same spot where their relatives grew a year or two prior. The two biggest plant families to look out for are the squash family, like pumpkin, watermelon, squash, and so on, and the tomato family like potatoes, tomatoes, peppers, eggplant, and so on. Crop rotation to various parts of the garden will help limit or prevent disease development and depletion of the soil's nutrients.

Dealing with weeds: Those pesky plants that seem to sprout from nowhere overnight are every gardener's pet peeve. Plan to weed your raised beds daily. Uprooting weeds is easier to do after watering or rainfall, but if the soil is muddy or wet, postpone the weeding until it dries out a little.

There are various ways to uproot a weed. One way is to gently pinch the base of the stem and tug out the root. Another way is levering out the root system using a weeding trowel. A hoe can also scrape off the top of the weed, taking care not to harm any crops. Remember that weeds will grow back if their roots are left in the soil.

Weeds are bad for your garden for many valid reasons. They not only compete with your crops for nutrients and water, but they also attract pests. If pests eventually find their way into your raised bed, they move from one plant to another, spreading disease. The best and most organic way to get rid of them is to pick them off by hand. If you hate insects or are squeamish, gloves might make you feel more comfortable.

Your garden should always be clean: Many diseases are spread rapidly through fallen, dead foliage. So inspect your raised bed at least once a week, or more if you can, to clean up shed leaves. It is possible to prevent a whole disease outbreak simply by getting rid of one infected leaf. Dispose of the diseased or dead leaves in a bin, never your compost pile.

Ensure that your plants are getting enough water and air: Watering your leaves in the afternoon or evening can promote the growth of mildews like downy or powdery mildew. Rather than water your crops from above, invest in a water-saving soaker hose that will deliver water straight to the roots, preventing splashing. Also, adhere to the spacing requirements on the seed packets to prevent crowding. Adequate airflow between crops can help prevent an outbreak of many fungal diseases.

Grow plants that attract beneficial insects: There are certain flowers that not only add beauty to your garden but also attract helpful insects like bumblebees that help with pollination and praying mantis and lady beetles that help devour harmful insects. Some of these flowers include:

1. Cleome
2. Daisy
3. Bachelor's button
4. Marigold
5. Purple coneflower
6. Cosmos
7. Nasturtium
8. Black-eyed Susan
9. Zinnia
10. Sunflower
11. Salvia
12. Yarrow

Chapter Seven: Vegetables for Raised Beds

When preparing for your raised bed garden, it is so easy to get overwhelmed by the many different plants in the seed catalogs. However, there are helpful tips to assist you in finding the right crops to grow.

Picking which plants to cultivate in your garden can be loads of fun, especially for beginners just itching to try everything. Looking through seed catalogs and marking out the seeds you are interested in cultivating is an exciting experience because the possibilities are endless. It is quite common to see beginner gardeners going overboard planting more crops than they need. Planning before planting is the smart route to take in order not to overwhelm yourself when all the harvests are ready at the same time.

How to Select Crops for Your Raised Bed Garden

This is time to think about what you intend to achieve with the garden. What, exactly, is the plan?

Do you aim to supplement your meals with fresh, organic produce?

Are you hoping to redirect the money you'd usually spend on groceries to something else?

Are you trying to avoid pesticides?

Do you plan to cultivate foods that can be preserved and stored through winter?

Chances are that the aim of planting your edible garden includes one or more of the points I will mention in this chapter. Having a clear picture of your objectives will help you better select the right crops that will be the most productive in your growing space. Here is a list of certain considerations when choosing crops for your raised bed garden:

Go for crops you and your family like eating: Unless you are an industrial gardener, it is pointless to put effort into tending to crops you or your family don't even like. Select crops based on your food preferences and that of your family. Gardening is especially satisfying when you are rewarded with the foods you like to eat.

If you enjoy salads, then greens, cabbages, lettuce, and tomatoes are great choices for your raised beds. If you like to have fresh salsa now and then, then onions, cilantro, peppers, and tomatoes should make the list. If you're not sure of the foods you enjoy, think for a minute about the foods you pick up often when you visit the farmer's market or produce section of the grocery store. What goes into the cart week after week? Yes, your garden might be small, but you will save a few bucks when you grow your own food.

Choose crops that thrive in your region: As mentioned earlier, seek information about the growing season and climate of your region. The best way to obtain this information is by speaking to other gardeners in the area. Suppose you have a gardener for a neighbor–even better. Talk to them about the crops they grow in their garden and the issues they face. Gardeners are naturally generous with their experiences and successful gardening tips.

High-value crops should make the list as well: What kind of vegetables do you enjoy eating but buy in little quantities and only when there is a sale? Do you see where I'm going with this? Growing crops that are usually expensive makes sense.

Common high-value crops are heirloom tomatoes, garlic, salad greens, and sweet bell peppers. For instance, a pack of organic lettuce goes for about $5 at the grocery store. But a pack of high-quality lettuce seeds go for less and will produce over six pounds of organic lettuce. Another class of high-value crops is herbs. During the growing season, cultivating an abundance of herbs to spice up your meals and preserve for winter is a good idea.

Consider replacing produce that has been contaminated with pesticide residue: The produce from some farms that is found in grocery stores has been contaminated with pesticide residue, whether consumers know this or not. Cultivating your own crops will guarantee you fresh organic produce, free of pesticides and other chemicals. This way, you get to eradicate completely or heavily reduce your intake of toxins.

A dirty dozen list is issued by the Environmental Working Group every year. This list contains the top twelve farm produce items that have tested positive for pesticide loads by the USDA. Crops that made the list this year include:

1. Spinach

2. Nectarines

3. Peaches

4. Strawberries

5. Cherries

6. Kale

7. Pears

8. Potatoes

9. Tomatoes

10. Celery

11. Apples

12. Grapes

These crops and more are easy to cultivate in your raised bed garden with zero chemicals.

Select crops that are easy to cultivate: If you are a beginner in the field or don't have enough time on your hands to tend to crops constantly, then consider cultivating crops that do not need to be coddled. One of the main reasons people give up on their gardening interests out of frustration is because they simply have no time for watering and weeding, which can be a lot of work. Certain crops don't require a lot of attention and care. Crops like artichokes, walking onions, asparagus, and many more.

Consider crops for preserving: If your goal is to preserve your harvest, you will have to plant enough to eat and to store. This can take a few seasons of experimenting to find your balance. Keep records of your experience and progress and adjust the number of crops every year.

There will be that time of year when your kitchen counter will be piled with tomatoes waiting to be transformed into salsas and tomato sauce. You will have carrots and string beans liking your refrigerator's crisper drawers eager to be pressure canned, and cucumbers ready to be turned into pickles. It can be overwhelming sometimes, but the end goal is always worth it. Therefore, it is important to consider yours and your family's nutritional needs first before selecting crops to plant. Also, think of the quantity of food consumed on average.

Note that certain vegetables like peppers, summer squashes, eggplants, and tomatoes grow and produce harvest throughout the growing season. Then there are others that only produce once, like carrots, garlic, onions, and radishes. Once you are done selecting the crops that you'd like in your raised beds, the next things to do are get organized, make lists, buy seeds, and get planting! Good planning is such an underrated aspect of gardening and is the key to a bountiful harvest. It doesn't matter if you are a beginner or have been gardening for years. You will benefit from planning every year.

Easy Vegetables to Plant in Your Raised Bed Garden

Many vegetables will thrive in a raised bed garden, but this is a list of the ones absolutely in love with the structured space provided by a raised bed.

1. Kale: This is one of the best greens to grow in a raised bed because it continues to produce well into the cool autumn season. With a raised bed, it is easy to cover the hardy kale with cold frames that will extend its productive season. You can use old box-framed windows to do this. If you live in a region with mild or short winters, you can ensure that your kales grow strong in winter with these raised beds.

2. Swiss chard: This vegetable revels in the supportive atmosphere created by the raised bed. They enjoy the loose soil and dense nutrients that ensure a lot of growth and tender, bright stalks. Grow this vegetable with kale and keep them producing well into the winter.

3. Carrots: So many gardeners suffer from terrible carrot harvest because they turn out stunted if grown in quickly packed-down soil; raised beds give them the perfectly loose soil they need to thrive.

Long carrots require tall, deep beds while tiny French carrots prefer short, low beds. Raised beds produce large and healthy carrots without all the stunted roots and knobs that conventional beds usually produce.

4. Parsnips: Like carrots, parsnips require soil packed with nutrients but loose enough to allow them to grow strong. These sweet roots will be grateful to be among the vegetables in your raised bed.

5. Tomatoes: Tomatoes are heavy feeders that germinate and spread to as many areas as they can. They are the perfect raised bed crop. They despise weeds and require a lot of attention and care to protect them from slugs and other insects. Construct a raised bed with four posters and use some twine to fence it in. This will provide space for your tomatoes to thrive.

6. Cucumbers: Cucumbers grow especially well in raised beds because they need soil with good drainage. When planted in the right conditions, it is safe to expect the crunchy, fresh, and tender cucumbers that are almost a dream to some. However, the cucumbers will harden if the soil stagnates. Their vines tend to take over the entire bed, so it is advised to place them in a separate bed and build them something to climb on.

7. Leeks: All onions thrive in raised beds because they are provided with well-drained soil and plenty of nitrogen, but leeks have taken a special liking to this method of gardening. Construct a low but long raised bed and grow your leeks as a stylish border in your garden. They don't take up as much space as tomatoes or cucumbers. They create a beautiful visual divider, giving you a full harvest of thick, tall leeks for autumn stews and soups.

8. Zucchinis: Every gardener knows that zucchini can sometimes be very overwhelming. It takes up the entire garden, producing more zucchinis than farmers need. Placing it in a raised bed will do nothing to reduce its yield, but it will prevent it from taking up all the space in your garden. Its spreading stems and wide leaves will be

given their own space. Don't attempt to plant anything else in the same raised bed as your zucchinis, with horseradish being the only exception.

9. Lettuce: Head lettuces make a beautiful addition to a raised bed garden. They are beautiful balls of dark green, bright green, or red-tipped green goodness. They also enjoy the warm loose soil and fewer weeds that come with raised beds. They can also be planted early in the season and keep producing until later in the season.

10. Beets: These roots are quite easy to grow. They love the loamy soil, but that isn't a problem for raised bed gardeners because literally, any soil can be purchased and piled into a raised bed. They can be planted all alone or share space with horseradish, another root crop that also enjoys pampering, fewer weeds, and friable soil. With raised beds, you can provide an environment perfectly designed for the beets, soil that is properly drained and contains the right amount of nitrogen. They also keep the horseradish in check, preventing them from spreading to other parts of the garden.

11. Salad vegetables: Cultivating salad greens like spinach and arugula in a raised bed is great, especially if you own dogs or chickens. Raised beds keep these greens inaccessible to intrusive pets, and the borders keep them well guarded against digging and scratching. Like lettuces, fragile salad greens also enjoy the warm soil and great drainage associated with raised beds.

12. Melons: This is another plant that will attempt to take over your garden! Place your melons in trellises in taller raised beds to contain them and protect them from slugs while they grow. Melons are slow-ripening crops, so they require a well-drained and more controlled environment or they are likely to rot on the damp earth. Cultivating melons in raised beds also gives you the chance to pay attention to the soil. The soil required to grow melons need to be consistently kept warm, and the raised bed environment is easier to manipulate than ground beds. Put your melons in raised beds over 6 feet high, and ensure that the soil is packed with organic matter.

13. Radishes: Radishes are assumed to be easy to grow, but the truth is that radishes are picky, moody little roots. They hate soggy conditions, dry conditions, over-rich soil, clay soil, and hot weather. They are the princesses of the plant world and will enjoy the lavish, controlled environment of a raised bed. Construct the perfect raised bed for your choosy radishes, throw in the perfect soil, and ensure it is well-drained but not dry. Do this, and you will be able to brag to everyone about the ease of growing radishes.

14. Potatoes: Mix your raised bed soil with lots of straw, and you will be pleasantly surprised at the ease with which your rootlets will grow into huge potatoes. These tubers are completely in love with everything about raised beds, which some people might find unexpected. Besides, potatoes are easier to harvest from raised beds because all you have to do is sit beside your bed and gently pull on the roots, instead of bending all the time.

15. Broccoli Raab: Broccoli can grow wherever space is provided for it, but broccoli raab is much smaller than the regular broccoli, so it easy for it to get overwhelmed by larger crops and lose water and nutrients when sharing space. Give your rapini its own home with an old wooden trunk. This tasty vegetable can be planted early in the season when the soil is warm, and because it grows rapidly, you can harvest multiple times before the season is over.

16. Celery: This is another picky crop that is practically demanding when grown in a raised bed. It requires rich, constantly damp soil and a long productive season. Your celery will show its appreciation for your attention by becoming more tender and tastier than you ever hoped.

17. Bok choy: A fast-growing, intense feeder, bok choy requires loose, rich soil. It doesn't appreciate sharing its space with weeds, which make it perfect for raised beds, especially in northern regions. It is a cool-weather crop that can keep growing well into the late fall, even without a lot of protection from the cold.

Chapter Eight: The Best Trees for Raised Beds

Small trees on decks or patios can add privacy, style, provide shade, act as natural focal points, and even produce fruit. The good thing is that a lot of these trees can grow well in raised beds or containers. Some possess special features like vivid fall colors, flowers, and attractive bark. However, some trees have messy features like dropping flowers, seeds, fruits, and others, and not everyone is comfortable with that. So, you must know all the features of the tree you intend to plant, and its survivability in your region.

Small Trees for Your Raised Bed

Here is a list of 13 amazing small trees to cultivate in your raised bed. Note: To select the perfect tree for your space, you need to consider its height and width at maturity. Also, some roots tend to crack or lift pavement, which would make it unsuited for a patio. If you plan to cultivate your tree in a pot, remember to check on it regularly to know when the roots need a new pot due to expansion.

Chaste Tree: This tree is an Asian and Mediterranean native with a lot of trunks that can be conditioned to create a nice shade tree. Its leaves are extraordinarily aromatic, producing tiny fragrant flowers on spikes during the fall and summer. The varieties latifolia and roses produce pink flowers, while silver spire and alba produce white blossoms, and it can be transformed into a shrub through pruning. Prune this tree every late winter to maintain its shape. The chaste tree is highly resistant to oak root fungus and is also heat-resistant.

Colors available: White, Pink, Lavender-blue.

Soil Requirements: Well-drained, loose soil, medium moisture.

Sun Exposure: Maximum sunlight.

Kumquat: Also called Citrus japonica, this tree can be cultivated in pots or on the ground. If they are planted on the ground, they can grow as big as 8 feet tall and 6 feet wide. The container versions are not as big but still as beautiful. Their bright orange flowers eventually transform into tangy edible fruit, and dark green leaves are a sight for sore eyes.

Plant kumquats in your raised bed for their bright orange fruits and aromatic blooms. They are a great addition to any garden, but they must be taken indoors during cold winters. Kumquats need to be relocated to a larger raised bed or container every two or three years and fertilized throughout the growing season.

Colors available: White.

Soil requirements: Wet clay or sandy loam.

Sun exposure: Maximum sunlight.

Japanese Maple: This tree is also called Acer palmatum. It is a naturally huge tree, growing up to 15 feet high at maturity. It can be cultivated on the ground and in raised beds. In raised beds, be prepared to transplant the tree to another bed every year due to a yearly increase in size.

There are different varieties of the Japanese maple, but the best ones for raised beds have finely cut, threadlike leaves and weeping branches. They include the Butterfly, Red Dragon, Crimson Queen, Mikawa Yatsubusa, Burgundy Lace, and Dissectum varieties. Japanese maples need not be pruned often. However, ensure that you rid the tree of damaged, diseased, or dead branches when you spot them.

Colors available: Red-purple.

Soil requirements: Slightly acidic, moist, rich, and well-drained.

Sun exposure: Maximum sunlight to partial shade.

Ficus: This tree is also called Ficus benjamina or weeping fig, and it can grow to a height of at least 50 feet in the wild, but when domesticated, it becomes a houseplant. It is a very eye-catching tree with its twisty, arching branches and bright green leaves.

Ficus makes a flexible patio plant that can transition easily from an indoor to an outdoor tree. It doesn't like the cold but can withstand the outdoors after the spring frost has passed. This tree requires monthly fertilization in the growing season but would prefer to be left alone during winter.

Colors available: Burgundy, green-purple, blue.

Soil Requirements: Well-drained, rich, and moist.

Sun exposure: Maximum sunlight to partial shade.

European Fan Palm: This tree, also called Chamaerops humilis, is perfect if you are looking to give your deck or patio a tropical vibe because the striking silhouette of the tree is an absolute beauty. There are also other species bred for small spaces including the paradise palm (Howea forsteriana), the Chinese fan palm (Livistona chinensis), the pygmy date palm (Phoenix roebelenii), the Lady palm (Rhapis excelsa), and the Windmill palm (Trachycarpus fortune). Always remember to fertilize your palm throughout the growing season and cut off any diseased or dead portions when they are spotted. Also, try not to overwater it because palms don't like that.

Colors available: Yellow.

Soil Requirements: Well-drained, rich, and slightly moist.

Sun exposure: Maximum sunlight to partial shade.

Ornamental Crabapple: This tree is also called a malus or flowering crabapple and is appreciated more for its short but worthy displays of pink, white and red flowers than its edible fruits. You can plant the smallest varieties in pots or containers while the others can be trained against a fence or wall as an espalier.

The varieties bred for large, raised beds are the Indian Magic, Sargent, Centurion, and Japanese, also called M. floribunda. Crabapple trees become more tolerant of the drought as they mature, but ensure that their soil doesn't dry out. If you experience long periods without rainfall, especially in the warmer months, water your tree. Also, they need to be pruned a little, outside the regular maintenance of removing diseased, dead, or damaged branches.

Colors available: White, red, and pink.

Soil Requirements: Well-drained, partly moist, and rich.

Sun exposure: Maximum sunlight.

Ornamental Plum or Cherry: This tree is sometimes called a flowering prunus. They are adorned with dark purple leaves and red, white, or pink flowers based on the variety. They can be planted in raised beds or large containers. A few varieties are susceptible to fungal disease and insect attacks, so ensure that you prune your tree to thin out the branches a little bit, improving air circulation, which helps to prevent these issues.

Small varieties of the flowering plums include Krauter Vesuvius purple leaf plum, also called Prunus cerasifera; Krauter Vesuvius; Double pink flowering plum, also called Prunus x blireana; and the purple leaf plum, also called Prunus cerasifera.

Small varieties of the flowering cherries include Yoshino cherry (Japanese flowering cherry), Okame (Prunus incisa x Prunus campanulata); Purple leaf sand cherry (Prunus x cisterna), and Albertii (Prunus padus).

Colors available: Red, white, and pink.

Soil requirements: Partly moist and well drained.

Sun exposure: Maximum sunlight to partial shade.

Pine: Also called Pinus, these evergreen trees provide you with something green to decorate your patio throughout the year. Plus, they provide a good amount of privacy and shade throughout the year. They like to be pruned frequently, so keep them as petite as you like. Certain species are bred for decks and patios like the Lacebark pine (Pinus bungeana), evergreen Japanese red pine (Pinus densiflora), and the evergreen Swiss stone pine (Pinus cembra).

For large, raised beds or containers, consider growing the Evergreen Japanese black pine (Pinus thunbergiana), and the evergreen mugo pine (Pinus mugo). Pine trees rarely need a lot of care. Simply water them during prolonged droughts and fertilize yearly if you have poor soil.

Colors available: Nonflowering.

Soil Requirements: Well-drained, fertile, and partly moist.

Sun exposure: Maximum sunlight to partial shade.

Smoke Tree: Also called the smoke bush, this tree is popular for its striking dark reddish-purple leaves and hairs so silky they look like puffs of smoke. It can be cultivated in a large container or raised bed, and near a patio or deck. The smoke effect is because of the fluffy hairs that accompany the tree's bloom in spring. The hairs transition from pink to purple as the summer progresses. Ensure that you prune very lightly in early spring to get the best blooms.

Colors available: Yellow.

Soil requirements: Well-drained and partly moist.

Sun exposure: Maximum sunlight.

Ornamental Pear: This tree is also called Pyrus. You will require at least two pear trees to cross-pollinate and produce fruit. If you have space for only one tree, choose between Bartlett or Anjou because they are the varieties with the ability to self-pollinate to an extent. Other small varieties for raised beds include Edgedell pear, Manchurian pear, Jack flowering pear, Snow pear, and Glen's Form. Pear trees have no issues with wet soil if they are provided with appropriate drainage. They are also prone to a disease known as fire blight, so you will need to prune off the diseased potions when they are identified to prevent the spread.

Colors available: White.

Soil requirements: Humus, well-drained, and moist.

Sun exposure: Maximum sunlight to partial shade.

Sweet Bay: This is also called Laurus nobilis. It is a tiny, slender evergreen shaped like a cone. Its leaves are highly aromatic and a dark green color. Its leaves are the exact bay leaves used for cooking many types of meals. It is a good choice for raised beds or containers placed on patios or decks. It can be pruned into a hedge or a topiary. It can tolerate drought but not for long periods, so water it when you experience long periods without rainfall. Yes, it loves to bask in a lot of light; however, shield it from the sun during hot afternoons, especially during warm months.

Colors available: Yellow-green.

Soil Requirements: Well-drained, rich, and moist.

Sun exposure: Maximum sunlight to partial shade.

Crepe Myrtle: Also called shrubs, these trees are extremely popular in the southern parts of the United States for their bright pink blooms, beautiful bark, and gorgeous fall leaves. You can plant the full-size varieties in large raised beds as they will grow as tall as 10 feet or choose from the many smaller varieties like Peppermint Lace, Zuni, Acoma, Hopi, Catawba, Chica Pink, Yuma, Pink Velour, Centennial, Seminole, White Chocolate, Glendora White, Chica Red, and Comanche. Try not to fertilize excessively because this can lead to excessive leaf growth. Excessive pruning is also unnecessary, although you can shape your tree early in spring if you wish.

Colors available: Pink and white.

Soil requirements: Partly moist and well-drained.

Sun exposure: Maximum sunlight.

Wisteria: Besides the obvious beauty of this tree, it can be conditioned as a shrub, small tree, or vine. If you want a tree, cut off all the stems, leaving only one and tying it to a wooden stake. When it has grown to the preferred height, pinch or prune the branch tips to force the growth of more branches. Wisteria can also be planted to layer a pergola or an arbor. The two most popular Wisteria species are Japanese Wisteria, also called W. floribunda, and the Chinese Wisteria, also called Wisteria sinensis. Don't use fertilizer unless your soil isn't rich enough, but feel free to layer a bit of compost to promote healthy growth and blooming.

Colors available: Purple, white, and pink.

Soil Requirements: Rich, moist, and well-drained.

Sun exposure: Maximum sunlight.

Chapter Nine: Raised Bed Herb Garden

People have used herbs for their healing and culinary properties for centuries. Today, herbs remain as useful and even more popular than ever. Chefs adore the unique flavors that herbs bring to all types of food and drink. Herbalists appreciate the healing properties of certain leaves, roots, and flowers. Herbal crafters preserve the fragrance and beauty of flowers and leaves in sachets, potpourri, dried arrangements, and wreaths. And gardeners love herbs for all their outstanding qualities, like their low maintenance, natural resistance to insects, and their vigor.

When many people think of herbs, it is common to picture the basic kitchen seasonings like rosemary, thyme, basil, and sage, but herbs are any plant that is deemed useful. For example, the seeds, flowers, leaves, roots, or stems of an herb can be highly valued for their medicine, dye, flavoring, fragrance, or some other benefit. It doesn't even have to be about function; many gardeners grow herbs simply because of how beautiful they are.

Where to Plant Herbs

Many herbs can survive in typical garden soil, if there is good enough drainage. Certain herbs like rosemary, bay, and lavender are woody plants with origins in the Mediterranean and thrive in sharply drained, gritty soil. Proper drainage is especially important because the roots of plants native to the Mediterranean tend to rot in moist soil. This is one reason why raised beds are perfect for cultivating herbs.

A lot of herbs flourish in full sunlight, enjoying at least six hours of direct sunlight daily. If your garden space doesn't receive as much sun, then consider herbs that don't require as much. Ideal choices include:

1. Parsley

2. Shiso

3. Mint

4. Cilantro

5. Chives

6. Tarragon

Like other crops, herbs can suffer when exposed to windy sites. Cultivating your herbs next to other buildings, walls, or next to your house provides for the warm and protective microclimate needed for them to thrive. It also increases your chances of a successful harvest with fragile perennials like rosemary. It doesn't matter if you grow the rosemary in a container and take it indoors during the winter, it is still recommended to spread it out in a sheltered but sunny area.

Where to Get Herbs for Planting

Certain herbs are fairly easy to begin from seed, but others take longer to germinate. Purchase slow-growing herbs at a nursery or simply divide existing plants if you have any. Some herbs can also be grown from cuttings.

Growing Herbs from Seed

Before planting any herb directly in your raised bed or seed-starting trays, go through the seed packet, which will help you with all the important information. Herbs that can easily be grown from seed include:

1. Borage
2. Chervil
3. Dill
4. Basil
5. Cilantro
6. Parsley
7. Calendula
8. Sage

Growing Herbs from Division

Perennial herbs can be easily divided. To do this, dig up the plant's root system using a garden fork and either use your hands to pull the roots apart or use a knife to cut the roots into as many pieces as you need, then replant them in your raised bed. You can also place a few divisions in pots to develop indoors well into the winter. If you plan to place the divisions outside, the ideal time to do this is fall. Herbs get established more quickly when divided and replanted in autumn. Perennials easily grown from division include:

1. Lovage

2. Oregano

3. Chives

4. Thyme

5. Bee balm

6. Marjoram

7. Garlic chives

Growing Herbs from Cutting

This method of growing herbs should be practiced in the summer or spring when plants are healthy, strong, and growing vigorously. Tarragon and rosemary have sturdy roots in the fall, which makes them great candidates for cuttings. Ideal choices for this method of growing herbs include:

1. Oregano

2. Thyme

3. Lavender

4. Sage

5. Mint

Steps to Grow Herbs from Cuttings

1. Pick stem portions that are tender and fresh, not woody. They also must be at least three inches long, with more than four leaves on them. Find a leaf node facing outwards and slash right above it with your knife.

2. Now pluck the leaves off the lower part of the stem and sprinkle rooting hormone powder all over the cut end.

3. Prepare a four-inch pot, filling it with wet potting soil. Now drive the stem about two inches deep into the pot.

4. Lightly cover the cuttings with a plastic bag because they need to be kept in moist conditions and away from intense sunlight. Avoid watering them until you absolutely must, and remove the covering if the area looks too moisturized.

5. Be on the lookout for fresh leaf sprouts in the first weeks because it means that the cuttings are properly rooted.

6. Time to move the newly sprouted plants into your raised beds filled with normal and healthy planting soil. This time, the raised bed must be placed under direct sunlight.

Growing Herbs in Planters and Pots

There are many advantages of cultivating herbs in planters and pots. They enable you to grow delicate perennials like flowering sages and rosemary throughout the year if you take them indoors during the fall.

1. Always begin with high-quality potting soil. This soil ensures good drainage. Avoid using normal garden soil because they tend not to have good drainage when placed in raised beds. Just like other plants in raised beds, herbs need to be regularly fertilized and watered throughout the growing season. Herbs native to the Mediterranean, like rosemary, have a high tolerance for partly dry soil, but only for short periods. Other herbs require more attention to watering, especially ones with broader leaves. During the growing season, when they are outdoors, use a liquid fertilizer according to the instructions on the package. When you bring them inside during winter, there won't be any need for much fertilization, except once or twice a month. Herbs can thrive in any well-drained, reasonably fertile soil, which makes raised beds or containers ideal for herb gardening.

2. Growing herbs need access to good lighting. If you can construct a hard path with bright-colored, reflecting paving, that would be great. Concrete or pebble panels are used in herb gardens

to reflect light onto the growing plants, creating a warm enough environment on chilly nights.

3. Herbs typically require little fertilizer and produce without a lot of feeding. Feeding can reduce the concentration of flavors.

4. A lot of herbs require soil with pH ranging from neutral to alkaline.

5. Intense levels of direct sunlight are crucial for producing good herb flavor, so your herbs should be placed in the most lit area of your garden.

Amazing Raised Bed Herbs

Here are my favorite raised bed herbs:

Basil: Also called Ocimum basilicum, this herb is a major ingredient in many recipes, particularly Mediterranean classics and summer salads. Basil is the most sold herb in all of Britain. Despite having origins in India, where it is regarded as sacred, it flourishes on British soil and is perfect for any raised bed.

How to cultivate: This delicate annual cannot withstand the cold and frost. It can only be cultivated outdoors during summer and must be taken indoors in colder months. It must be grown in fertile soil and be exposed to as much light and warmth as possible. Greenhouses are perfect, same as kitchen windowsills, to keep basil thriving for a long period. There are many available varieties, so experiment with as many as you like this summer and relish the tastes on your pasta dishes and homemade salads.

Chives: Also called Allium schoenoprasum, this hardy perennial is especially easy to cultivate. Chives are a great addition to your herb garden as they are popular for their beautiful purple blossoms. Once upon a time, they were once strung up in bunches to ward off evil spirits, but they are a major kitchen ingredient today. All parts of the plant are edible, which makes it extremely versatile. Its flowers can be used to garnish various salads while the leaves and bulbs can add flavor to many meals. Having a light onion flavor, chives come in handy when preparing all kinds of summer dishes like omelets, classic potato salad, and soups.

How to cultivate: Chives are very low maintenance. You need only to plant them in soil and ensure that they are kept in a sunny spot where they can receive direct sunlight for at least five hours.

Mint: This is also known as common mint, spearmint, or Mentha spicata. It is a hardy herb that can be easily cultivated in any garden. It flowers light purple blossoms throughout August and September. Being a perennial, mint can be trusted to grace your garden year after year. It is vigorous by nature, so don't be surprised if it invades other parts of your garden. To avoid this, plant it in a bottomless raised bed set on the ground. Its pleasant and refreshing spearmint flavor is often used to spice up salads and sauces. Mint leaves are also dried and used to prepare fresh herbal tea or domestic herbal medicines.

How to cultivate: Mint only requires fertile, moist soil and direct sunlight. It is very adaptable and can thrive in most situations; it is not prone to frost damage.

Coriander: This is also called Coriandrum sativum or Chinese parsley. It is a delicate and short-lived annual, which is only grown from seeds planted at intervals throughout the growing season. The whole plant can be eaten; it is very popular in culinary culture and is typically used in Asian dishes, including Chinese and Thai meals.

Its leaves and seeds have very distinct flavors. The seeds taste a little like lemons and can be crushed to be used as a spice. The leaves are more bitter and can be chopped up and used to garnish meals. Besides its many culinary uses, it has many health benefits as well and is a major ingredient in herbal remedies worldwide.

How to cultivate: Coriander enjoys fertile soil and adequate sunlight. It prefers partial shade as this prevents the seeds from setting prematurely.

Dill: This herb is also called Anethum graveolens, and it is a hardy but short-lived annual. It is relatively easy to grow from seed in your raised bed, and can be used for a variety of things like cooking and production of certain cosmetics. Dried or fresh dill leaves with their wonderful fragrance combine beautifully with seafood like smoked salmon. It is also popularly paired with soups and potatoes.

How to cultivate: Grow in moist soil where enough warmth can surround the herb. Partial shade is preferred to prevent the premature setting of the seed.

Fennel: This indigenous Mediterranean herb, also called Foeniculum vulgare, will make a lovely addition to your herb garden. Also native to the Mediterranean, it can be cultivated from seed in the UK. Despite being a hardy perennial, fennel is usually grown annually to maintain its crop. It is highly aromatic with an aniseed flavor, which makes it an amazing ingredient for both savory and sweet dishes. Its young, delicate leaves can be used as a garnish in soups, salads, and with seafood sauce, as well as in sticky, sweet, delicious puddings and sauces. The whole plant is edible, making it a versatile herb.

How to cultivate: This is an especially robust herb that grows well in any soil if it is placed under direct sunlight.

French Tarragon: The French tarragon is also called Artemisia dracunculus, and despite being a little tricky to cultivate, this herb is loved by culinary enthusiasts, particularly those of the French cuisine. It has a sweet anise aroma and licorice flavor. It is considered the finest of all tarragon varieties in the kitchen. It is especially delicious when combined with chicken, but it is also used to season oils, vinegar, and béarnaise sauce.

How to cultivate: Despite being perennial, it is prone to rotting out in over-saturated soils and wet regions, so take care to plant in partially dry soil and not water excessively. Cultivate in fertile soil where it can receive adequate amounts of sunlight and warmth to produce shoots in abundance.

The French tarragon rarely blossoms, so seed production is greatly limited. It cannot be cultivated from seed and must be grown through root division. Divide the roots in spring to maintain its health, and replant every two of three years.

Parsley: This popular herb is also called Petroselinum crispum, and is an absolute must in your herb garden. It is a hardy biennial grown from seed every year in summer and spring.

It is used to prepare Middle Eastern salads, pesto paired with basil, and used in fishcakes and stews. Curly parsley is very decorative due to its curly leaves and is used to garnish many dishes.

How to cultivate: For the most productive results, plant this in the fertile soil of your vegetable raised bed. Water regularly during prolonged periods of drought. Parsley tolerates a little shade, although it loves direct sunlight. There are two kinds of parsley cultivated in Europe: The flat-leaf and curly parsley. The flat-leaf is more popular because it is more tolerant of sunshine and rain, and has a stronger flavor, according to some.

Rosemary: This herb is also known as Rosemarinus officinalis, and it is considered great for brain health. It is believed by the Greeks to be linked to having a good memory and cognitive function. It is an especially nutritious herb to cultivate in your herb garden. Being an evergreen shrub, it is available all year round and has aromatic leaves that are beautifully shaped like needles to adorn your garden. Rosemary is also considered a decorative plant because of its white, purple, and pink flowers. Combine rosemary with roast meats such as chicken and lamb, and use it as a flavoring in Yorkshire puddings and stuffing.

How to cultivate: This herb thrives in well-drained soil, under direct sunlight. It is resistant to pests and tolerant to dry spells, but not for long periods.

Sage: Also called Salvia officinalis, this herb is known for its intense flavor and slightly peppery and savory taste, making it one of the most widely grown and used herbs in Britain. Its green and white and purple variegated forms make it an exceptional source of adornment for herb gardens, particularly because it can also act as an ornamental border. This kitchen essential is often used in stuffing and paired with pork.

The usual property of this herb is the significant increase in flavor as the leaves grow, so the bigger the leaves, the tastier the dish. Besides being a good source of Vitamin C, it is rich in other minerals like potassium.

How to cultivate: This evergreen shrub will be available to you year-round if it is cultivated in well-drained areas with lots of sunlight.

How to Dry Herbs

Airdrying

1. Select about five to ten branches and hold them together with a rubber band or string. The fewer the branches, the quicker they will dry.

2. Place the bundle of herbs in a paper bag with the stem side facing up. Use a string to seal the bag closed, ensuring that you don't crush any herbs. Now make a few holes in the bag for air circulation.

3. String the bag up by the stem end in a warm, well-ventilated area.

4. Your herbs should be dried and ready for storage in a week.

Oven-Drying

1. Spread the herbs on a cookie sheet with a depth of 1 inch or less.

2. Slide the sheet pan into an open oven and let it dry on low heat for two to four hours.

3. To test if the herbs are dry enough, touch the leaves to see if they crumble easily. Herbs dried in the oven tend to cook a bit during the process, which removes some of the flavor and potency, so you might need twice the normal amount when using them for cooking.

4. Preserve the herbs in airtight containers like plastic storage cabinets, freezer Ziplock bags, and canning jars. For the perfect flavor, don't crush the leaves until you are prepared to use them. Also, be sure to exhaust them in a year.

Freezing

Certain herbs retain their flavor best when frozen. They include dill, basil, lemon balm, chervil, mint, chives, rosemary, thyme, parsley, lemon verbena, French tarragon, sage, and oregano.

1. Wash the herbs properly and pat or shake to get rid of the excess water. You can chop them before storing them if you like.

2. Put the whole or chopped leaves in freezer bags and flatten them to get rid of air.

Rosemary, thyme, dill, and sage freeze well on their stalks, which can be added to cooking pots frozen and removed before serving. You can also blend the herbs into a puree with a little amount of water and freeze the paste in freezer bags. When you are ready to use them, simply chip off the frozen pieces and throw them into your pot.

Mix herbs that are good cooking companions such as thyme and sage. Throw them in a blender with a drizzle of olive oil and puree until smooth paste forms. Pour the mixture directly into freezer bags or ice cube trays and then freezer bags.

Chapter Ten: Growing Flowers in Raised Beds

The most vital reason to plant some flowers in your raised beds is to draw in native bees and other pollinators. If bees don't make a pit stop at your garden for a quick nectar snack and to throw some pollen around, you will be pretty disappointed with your crops. Besides, cultivating bee-friendly flowers in your garden supports biodiversity and struggling pollinator colonies. There are so many flowers particularly designed to attract hummingbirds, butterflies, and other nectar-loving species.

Before you make any seed purchases, here are a few important tips to remember when choosing flower varieties for your raised bed garden.

• Take note of bloom time: To succeed in companion planting with flowers, you need to choose flowers with the same bloom time as your vegetable crops. If the flowers you settle on do not bloom until three weeks after your peas are done flowering, your peas will not benefit from that companionship. Seed packets will give you the necessary information about the flower, including its bloom time, so you can synchronize your planting schedule. Grow a variety of flowers with different bloom times to ensure that all or most of your veggies benefit from the experience.

• Consider the flower shape: The flowers that draw in beneficial wasps or bees are not the same kind of flowers that pull in hummingbirds. The shape of the flower makes access to nectar and pollen harder or easier for different species. If you are looking to attract pollinators like bees, consider flowers with a composite shape like daisies, purple coneflowers, zinnias, sunflowers, and cosmos.

• Spread them out: Don't plant all your flowers in one section of the garden; space them out. How you do this is your decision, but there are many ways to go about it. You can cultivate a row of flowers right after a row of vegetables, or you can plant one flower between two vegetables. Come up with your own strategies, like using flowers as a border or to break up a row as an indicator of where a certain vegetable ends and others begin.

• Consider the height of the flower: You don't want flowers that will compete with your crops for sunlight so go for primarily low-growing flowers. For instance, certain crops like spinach might appreciate a little shade during the warmer months, so the height of your flowers depends on their location and the crops in your garden.

• Start with simple flowers: If you are a beginner, I recommend that you begin with annuals because they are easy to grow and produce a lot of blooms. Also, you won't need to worry about them sprouting in the same spot next year if you intend to change the design of your garden. One of the most effective ways to attract native bees is to plant native perennials, so plant them in small amounts; you know what will work best for your garden.

Quick Glossary

1. **Perennial:** Any plant that flowers year after year. The leaves typically fall to the ground in autumn, die, and then regrow the next season. Some perennials last longer than others like peony rose, while others don't last past a few years.

2. **Annual:** Any plant with a growth cycle of one year. Annuals are typically hardy crops and remain unaffected by frost. A great example is calendula.

3. **Half-Hardy Annual:** Any annual that is delicate. They are usually cultivated indoors or in warm areas like greenhouses, then cautiously exposed to cold conditions outdoors now and then before being transplanted outdoors after the threat of frost is over. For example, you have marigolds, cosmos, and so on.

4. **Biennial:** Any plant with a growth cycle of two years. Biennial seeds are usually sown around April of the first year. The leaves mature the same year and then flower the next year. Examples are wallflowers, foxgloves, and so on.

20 Best Flowers for Raised Beds

- **Himalayan Blue Poppy:** This is a perennial with gorgeous sky-blue color. It is a half-hardy crop that enjoys little moisture. Remove the flower buds in the first year to prevent them from blooming; otherwise, they may become reluctant to bloom again.

- **Annual Poppies**: These are relatively easy to plant from seed. They produce a lot of seeds in late summer and early fall. The seeds can be harvested and stored in a cool, dry place like a paper bag until next season when they can be scattered to sprout new plants.

- **Aubrieta:** This perennial is considered a ground covering plant, making it suitable for raised beds and walls. It can sometimes sprout from cracks in the wall and pavement, spreading several feet away from its origin. It can be found in pink, rose, lilac, and purple colors. This flower can last for as long as ten years or more.

- **Red Valerian:** This flower is available in white, red, and pink colors. It is commonly found growing in the wild, ruins, and on old walls and bridges. It can spread throughout your garden and become a weed if left out of control. It is a very hardy plant and can withstand extremely cold temperatures.

- **Delphinium:** These perennials are tall and have spikes. They have sky-blue, violet, and white variations. They can withstand low temperatures with no form of damage to their roots, but they must be staked to prevent them from being flattened or blown away by heavy winds. As perennials, they can last for as long as twenty years. They also produce seeds that can be stored and easily planted the next year.

- **Geraniums:** These are the same as pelargoniums, which are also called geraniums. These herbaceous perennials come in a variety of beautiful colors. They are capable of self-seeding and spreading all over your garden if not controlled.

- **Scabiosa:** This is a perennial that has cream, lavender-blue, and lilac varieties. They require little attention, but you must remove the dead heads for new buds to form. It has a relative tolerance of low temperatures.

- **Perennial Flax:** Also called blue flax, this plant has beautiful sky-blue flowers adorning slim, tender stems. Their flowers last only a day but are constantly replaced by new ones. It is one of the short-lived perennials.

- **Livingstone Daisies:** Also called mesembryanthemum, these flowers are the low-growing kind. They are also annuals and quite easy to plant from seed. When they bloom, the flowers are multi-petaled with a variety of colors. The petals open in the daytime and close at night.

- **Bellflower:** Also called Campanula, this ground-covering plant is a low-growing perennial with lilac flowers shaped like a small bell. It grows rapidly and spreads over rocks and walls.

- **Rockrose:** These are sub-shrubs that grow rapidly and are great for edging. They thrive in direct sunlight and well-drained soil.

- **Silver Ragwort:** This biennial is also called the dusty miller. It is a Mediterranean native popular for its ornamental silver-colored leaves. It is relatively hardy and can survive hot temperatures with little scorch damage. Being a biennial, the seeds are typically planted in April on trays and transplanted to raised beds or pots once they sprout their first pairs of true leaves. Like all biennials, it blooms in the summer of the next year. It is also toxic and must be kept out of reach of livestock and children.

- **Foxgloves:** These are biennials just like silver ragwort, so the planting process is the same. Sow the seeds in April, transplant in the fall, and then watch them bloom in the summer of next year. Environmental regulations in some regions ban digging up of plants from the wild and replanting in your garden. However, this biennial is quite easy to grow from seed. They are relatively tall plants with spikes and purple flowers shaped like bells.

- **Oriental Poppies:** These are relatively easy to grow from seed. They are typically sown in spring and germinated near a hot water tank or in a warm room of 17°C. Spread the seed on a tray layered with compost and sprinkle some extra compost on the top to sow. They must be transplanted into raised beds or pots once they are sufficiently large.

• **Feverfew:** This perennial herb is adorned with white daisy-like blooms. It self-seeds readily and will spread to all parts of your garden if you don't keep it under control. Some people are not very fond of white flowers, but a splash of white here and there can balance out the striking colors in your garden. This plant is popular for its fast action on migraines if the leaves are infused into hot water to make an herbal tea or chewed raw, although I don't recommend that because it is BITTER.

• **Oregano:** This perennial herb is known for its subtle but beautiful flowers. Its stems are adorned with little pale pink flowers when it blooms throughout summer. It is a hardy shrub capable of colonizing your garden if you let it. It is also a favorite of bees, butterflies, and chefs!

• **Pot Marigolds:** These are also called calendulas. They are hardy annuals quite easy to grow from seeds. They readily self-seed and are available in yellow and orange varieties.

• **Hyacinths:** Flowers usually have no other choice but to brighten up the garden in spring, but hyacinths arrive late to the party with their splashes of color in late spring and early summer. They are grown from bulbs and are typically sown in autumn to bloom in March and April. They are naturally hardy but can be conditioned to be delicate if cultivated in pots. When shopping for hyacinth bulbs, buy the ones with a 'prepared' label on them. This variety needs to be forced and should be sown in late September in a cool place with a temperature of about 50 degrees Fahrenheit for about eight to ten weeks. Then they are ready to be placed in indoor raised beds to flower in approximately three weeks. Hyacinths come in varieties of white, purple, and various shades of blue.

• **Red Campion:** This perennial is a wild one and breathtakingly beautiful. It is available in varieties of pink and red and, like many wildflowers, red campion isn't as flamboyant as the flashy bedding plants popularly used in planters. When paired with other wild plants to form a border, the result is more natural and subtle. The individual seeds are generally available for purchase, but if not, ask your seed provider if they can be added to a packet of different wildflower seeds.

Chapter Eleven: Preparing Your Beds for Next Year

Fall is finally here and, as expected, it brings with it the inevitable slowing of activity in every garden. Based on your location, perennials may have started blushing with beautiful colors and shedding their leaves.

Annual veggies are reaching the end of their lifespan and are beginning to yield to the nip of progressively heavier frosts. Following the rush of spring sowing and the peak of summer's reaping, it is tempting to pull the curtain, sit back, and just let nature do its thing. You already did the heavy lifting in spring and reaped the benefits in summer. What else is there to do now that fall is here?

The answer to that question depends on how smoothly you'd like to transition into spring when it rolls around. A few cautious steps executed this season will save you a lot of time and effort in the long haul. If you'd rather lessen the work that comes with the start of next year's growing season, then take some of these suggestions into consideration. Let's look at the steps to put your garden to bed:

1. Gather up and dispose of finished and rotting plants: besides the unattractive look they add to your garden, old plants can harbor fungi, diseases, and pests. As noted by the Cooperative Extension of Colorado State University, unwanted pests and insects who feed on your plants throughout summer may deposit eggs on their leaves and stalks. Getting rid of spent plants from the surface of the soil keeps pests from getting a head start when spring rolls around. You can also bury them in garden trenches if they are free of diseases, as this improves soil health by contributing organic matter to the soil.

2. Get rid of invasive weeds that may have spread over the growing season: Remember the intruding Himalayan blackberry? Or the bindweed that took over your raspberry patch? The time to get rid of those renegades is now. Pull them out of the soil with their roots and trash or burn them. A lot of invasive weeds remain active in weed piles or compost heaps, so don't give in to the urge to move them to another section of your garden. Getting rid of invasive plants is the only way to keep them from growing all over again and being their disruptive selves when spring comes back.

3. Your soil needs to be prepared for spring: Even though many gardeners would rather perform this activity when spring comes around, fall is a good time for soil preparation like adding manure, bone meal, rock phosphate, compost, and kelp. In most regions, the climate allows these new additions to break down, enrich your soil, and become biologically active. It also means there will be no need to wait until your garden is dry enough in spring to be worked on for the first time.

Turning, digging, and amending your soil now gives you some time off when the season hits because you will have already done most of the work. Also, tilling in fall helps boost soil drainage before intense weather becomes a reality. Once you have made all the necessary amendments according to your soil needs in fall, layer the bed with sheet plastic or any other safe covering to keep winter rains from sinking the newly added nutrients below the active root zone.

This applies to all types of gardens but especially to raised beds because they drain more effectively than ground beds. Take the sheeting off in early spring and use a hoe to till lightly before planting.

4. Consider planting cover crops: In many climates, early fall or late summer is a great time to plant cover crops like clover, rye, or vetch. These crops help protect your soil from erosion, boost levels of organic matter, and break up firmly packed areas. Cover crops also contribute to the nutrient content of the soil. Planting legumes like field peas or clover in your raised beds helps to enhance your soil's nitrogen levels, which is an important aspect of vegetable gardening. It is recommended to plant cover crops about a month or more before the first killing frost hits, but certain cover crops are stronger than others, so check with your seed provider or local extension officer to know the best cover crops for your region.

5. Prune your perennials: Fall is a great time to prune perennial garden crops, although you should be cautious when choosing the ones to prune. For instance, fennel likes to be pruned in the fall. Research has shown that dead raspberry canes continue to feed the plant's crown well into the winter. Blueberries also like to be pruned in spring as it helps protect them from stress and disease. Direct your fall pruning efforts to herbs like thyme, sage, and rosemary, and vegetables like rhubarb and asparagus. Blackberries also like to be cleaned up nicely in the fall. Get rid of any crossing or spent canes to help control the plant's aggressive spread.

6. Consider dividing and planting bulbs: Despite the flowering and death of spring bulbs, other flowering bulbs like lilies bloom in fall. Wait until three or four weeks after the bloom to dig up and divide any crowded or straggly plants during this season. Lift the bulbs carefully and divide bulblets to be transplanted immediately to another section of the garden. If you dug up your spring bulbs before fall rolled around, now is a great time to replant them.

Crocuses, daffodils, and tulips should be ready to get back into the soil for another glorious display next season.

7. Harvest your compost pile: Now that the summer heat is done and the microbes of nature are getting ready for their winter nap, it is tempting to ignore the compost pile, but I'll tell you the two ways this can be a missed opportunity. First, materials left to decompose over the summer are done and ready for use. This rich material should be layered on garden beds to fix deficient soil, fertilize lawns if any, or generally top up the garden bed. This will nourish your plants and give their growth a jump start when spring comes around. Second, clearing out your compost heap will make way for a fresh batch, which can be insulated against frost, meaning that microbes can get to work even in the winter. Pile up lots of autumn leaves, sawdust, or straw, and layer them with scraps from the kitchen and any other active green matter you can lay your hands on, to keep those microbes working for longer on your fresh compost pile.

8. Replenish the mulch: Mulching in winter has a lot of similar benefits with mulching in summer. They include a significant reduction in water loss, protection from erosion, and prevention of weeds. Mulching in the winter has more benefits because as the soil transitions to colder temperatures, the freezing and melting of the soil can have adverse effects on the crops whose roots have to go through all the churning and heaving associated with the transition. Layering the soil with a mulch helps with temperature and moisture regulation, which eases these effects. Piling a thick layer of mulch around the root crops left in your raised beds this season can act as a buffer against killer frosts and prolong the life of your crops. Plus, the mulch decomposes to add fresh organic material to your soil.

9. Assess your growing season and review the cultivars: Did your selection of fruits and vegetables perform as well as you hoped this season? The time to reconsider under-performing crops is now. You need to take stock and discover if there are plants to be replaced, and if better varieties are available for your location. If

your crops did as well as you hoped, I suggest extending the harvest by adding varieties that ripen later or earlier in the season. When taking stock of vegetable performance, take cautious notes about what worked and what didn't to prepare for next season. Some failures and successes of the season can be because of the climate, but other factors can be controlled like moisture levels, plant orientation, and soil fertility. Keep a record of the lessons you learned this season, the highs and the lows of summer, as they will act as a reference for the next planting season.

10. Care for your tools: Taking care of your tools is a given in gardening, and most gardeners know this. This crucial task can seem overwhelming when the farming season is back, and there is just so much to do. Fall is the perfect time to show your tools some love and affection. Start by washing them to get rid of dirt and debris. If a tool is rusty, file it with a wire brush or sandpaper. Use a basic mill file to sharpen shovels and hoes. Use a whetstone to sharpen pruning shears and other bladed tools. Finally, clean all the tools with a rag lightly coated in machine oil. This helps protect the metal from oxygen and extend the lifespan of your tools for the next season.

The Importance of Foresight

Regardless of where you live, there will always be steps to take to prepare for the next gardening season, as outlined above. When you take these steps, they will not only ensure a smooth transition into the next farming season; they will also boost your harvests in the long-term.

Preparing for a New Raised Bed

If you plan on purchasing new raised beds next season, the best thing to do is get them ready the fall before. As we discover more information about soil health, especially the economically important microbes that live in the soil, we realize the importance of having a head start on certain processes before the plants are ready to be cultivated. This way, they get settled in and established more quickly than usual, and remain healthy for as long as those conditions are maintained. Amending and tilling the soil long before any planting allows the soil to come alive over the winter season. Some gardeners like to call this process "building a living soil profile," and it needs some time to activate. This process can be done with ease and involve a few simple steps.

First, select a location for your new bed and mark it off with a barrier. Once that is done, use a shovel or tiller to turn the soil over. Now pour in your compost. If you have domestic animals, their poop can serve as composted manure, but any good compost will suffice. Some people make theirs; others make purchases. If you want to go down that path, as long as it isn't sterilized, you're good to go.

Another great source of compost is the pile of leaves being shed this time of year. If you have a mower fitted with a bag, shred the leaves before using them. If not, just pour them directly on the bed whether they are fresh or not, because they will be decomposed by the time its farming season. Those important microbes we talked about will get to work on them.

Don't manage your compost. Pile it up as high as 4 to 6 inches, especially if it's free. Begin by layering the surface of the soil and then work it into the soil with either a tiller or shovel. When that is all done, get a hand rake and level your garden bed.

The next thing to do is absolutely nothing. Just sit back and let nature take the wheel. What will happen next is those important microbes will come alive and grow inside your soil. To give the process a jumpstart and take things up a few notches, you can throw in a microbial drench. These are packed with a crazy amount of microbes, and all you must do is water them into your composted soil. They will begin to eat the compost, breaking it down in the process and turning it into nutrients that will be eaten by your plants when the spring rolls around.

The other amazing part about setting up a bed now is that garden chores are usually at a minimum during the fall. This gives you ample time to do it right without having to worry about watering, fertilizing, or weeding other raised beds. The key to a healthy and productive garden is a solid foundation, and that foundation is your garden bed. Take the right preparatory steps, and your gardening will be a lot easier and more productive.

Chapter Twelve: The Importance of Charting Your Progress

Keeping a record of your progress and the happenings in your garden is something most gardeners start enthusiastically but eventually let fall by the wayside. When the crops start growing and you begin to reap the harvests, it is easy to forget to take notes. The usefulness of the information and photos taken during the growing season is so underrated. With a journal, you can look at things in retrospect, seeing the problems you had and the time they occurred, the plants that flourished, and the ones that didn't.

Here, I will outline an easy method for keeping track of the things you planted and where you planted them. All your questions might not necessarily be answered, but it will give you a head start. Let's start with the supplies you will need to begin your garden journal:

1. Three-Ring Binder: Any binder will suffice, but if you intend to take your journal into your garden now and then, a vinyl binder is a perfect option. Get a binder that zips shut, making it easy to slip things into it even when you're in a hurry and not having to worry

about them slipping out. You can also get one with a cover that lets you slide a photo of your favorite container, plant, or raised bed.

2. Plastic Photo Sleeves: Plastic sheets are relatively cheap and are sold at craft stores at a discount. Sleeves the size of baseball cards are also available and sold in bulk, but they are too small for some seed packets, so get them in a variety of sizes for seed packets, photos, and tags.

3. Blank Pages: Put blank pages in the back of your journal for extra notes. If you make notes somewhere else for some reason, you can always slide them into a sleeve.

4. Permanent Markers: You will need at least one of these for longer notes.

5. Seed Packets and Plant Tags: Store them as you plant.

6. Calendar: This will enable you to keep track of your planting days and any other event significant to your garden.

7. Pictures of your Garden: Ensure you take photos of the good and bad. Most of the time, we don't think to take pictures of insect damage and diseases, but they are just as important as the blooming flowers.

Starting Your Garden Journal

Once you have gathered all your journal supplies, now you must tuck things into your plastic sleeves. Preserve all your seed packets and plant tags by slipping them into the sleeve pockets. For convenience, keep information about each raised bed or garden section in a separate sleeve to make it easier to find information about your plants based on their location.

Now that you have all that information in one place, the next course of action is note-taking. The packets and tags give you information on your plants instantly. Both sides of a sleeve are see-through, so you'll be able to easily obtain plant identification and the notes on their development.

Your permanent markers will help you make more notes. You can start by marking the raised bed or area of the garden directly on the plastic sleeve with distinct colors. You can also take notes of all the details like when a plant first bloomed, the harvest you obtained, the day you planted a crop, and where it was purchased.

All you need to get started on this is a fine tip marker. You can get more creative with stickers, glitter, markers, and other items from the craft store if you wish. If you have a calendar, it will serve as a handy reference for all the significant dates like the day crops were planted, pruned, and harvested, even when they bloomed or wilted.

Taking Photos for Your Journal

Taking photos for documentation is not only an amazing way to keep track of your garden's progress, but it also acts as a trigger for your memory years or even months after you began keeping records.

As your garden progresses and your plants are sprouting, take and print shots of your favorite scenes and great combinations. Shoot the same spot at different times, on different days, to observe the progression. Take photos of the raised beds or containers you would like to duplicate. Try using the black and white picture trick where you observe photos without the distraction of color. Pictures are also a nice way to take notes about crops that need to be moved or divided and colors that are striking or are too dull.

Also, make sure that you take shots of diseased crops and pest infestations to keep track of the issue. Endeavor also to take pictures of any area with issues, no matter what they are, to study them and correct the problem during the off-season. Don't worry; no one else has to see them but you.

Making Use of Your Garden Journal

Now that you have all the necessary tips, it is time to put them to work. Typically, you will have to take your journal with you into the garden to take quick notes, but having all the pictures, plant tags, and seed packets will enable you to go through your garden from the comfort of your study or bedroom and jot down all the important notes and reminders

Specific Tracking Suggestions

1. The crop you planted and where you planted it.

2. When you began seeds.

3. Sections of the garden that need to be worked on.

4. Issues to closely observe.

5. Crops that need to be moved or divided.

6. Overgrown areas of your garden.

7. Crops that require attention in spring.

8. Where crops were planted in preparation for crop rotation and much more.

As time passes, you will discover many other uses for your garden journal, and you will be grateful for keeping one when the time comes to purchase more seeds or plants in preparation for the next season. And if you ever relocate, it will be easy for you to compile a plant list with the necessary information for the next owner of the property–assuming your raised bed is fixed, and the new owners are into gardening.

If you keep an ornamental garden, making notes is equally useful, especially as you try out new plants every season. If you'd rather tend to the same plants every season, I suggest a perpetual journal. It is similar to the normal journal because you will still store plant tags and seed packets in sleeves, but you can also keep a running record of your garden's progress every year. Do this and watch your journal evolve to suit your gardening style.

Conclusion

Raised beds have been around longer than the word itself. Since they are simply gardens where the soil is elevated from the ground, other advantages besides aesthetics might not be obvious to many people, well, except people with bad backs who see the convenience at first glance.

You don't need a raised bed to cultivate great-tasting fruits, vegetables, and herbs because almost any ground bed with maximum sunlight can do that. However, raised bed gardening stands out because of its numerous advantages you are now aware of. For one thing, it is incomparably easier on the back, as it involves less bending over.

Raised beds are your opportunity to start over with uncontaminated and enriched soil. Suppose you lived on sloped property; no worries because raised beds offer level and convenient planting areas. Plus, they warm up faster than ground beds in the spring, so you get a head start when it's growing season.

The numerous advantages will not do much for you if you neglect your soil, and that is one of the most common mistakes made by beginning gardeners. When your soil is healthy and packed with nutrients and organic matter, your crops will be more robust and practically care for themselves. With raised beds, you get to water and weed less and worry less about pests and insect attacks. With raised beds, you call the shots.

This book has provided practical advice on the kind of mulches and frame materials to use, different ways to boost soil fertility, seed selection, and lots of options to choose from including the different options for irrigation, pest prevention tactics, and much more. You have all you need to know to get started on raised bed gardening. It is the perfect blend of convenience and productivity, and I know that you will have as much fun gardening as you hoped!

Here's another book by Dion Rosser that you might like

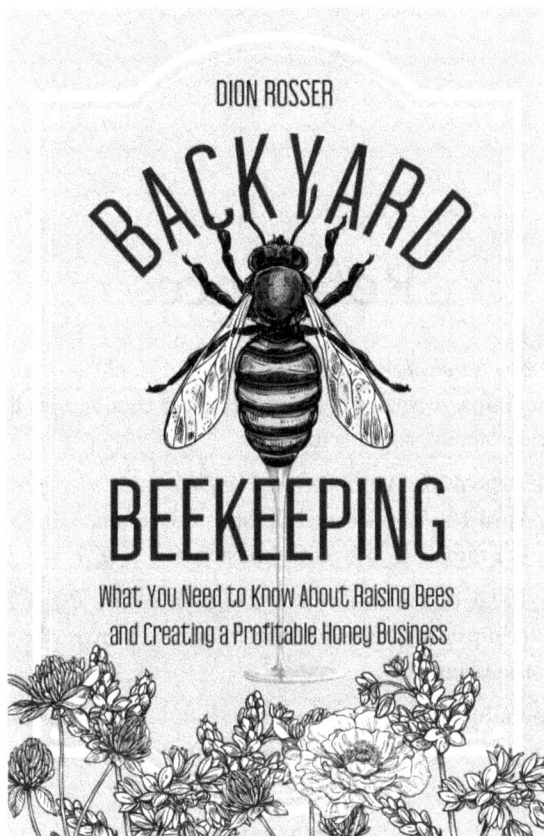

References

10 off-grid homes for a self-sufficient lifestyle. (2019, February 16). Dezeen. https://www.dezeen.com/2019/02/16/off-grid-self-sufficient-homes-sustainable-architecture/

103 Ways to Make Money from Your Homestead, even a small one. (2019, April 16). The Rustic Elk. https://www.therusticelk.com/103-ways-make-money-from-your-homestead/

Aaron, P. (2018, December 20). *Off Grid House Plans.* The Simple Prepper. https://www.thesimpleprepper.com/prepper/homesteading/off-grid-house-plans/

admin. (n.d.). *Off Grid Living.* Http://Off-Grid-Living.com. Retrieved from https://off-grid-living.com/getting-started-off-grid/

All About Off Grid Wastewater: Options, Septic, Code, and Advice. (2017, July 25). Accidental Hippies. https://accidentalhippies.com/2017/07/25/off-grid-waste-septic/

Beckett, T. (2019, May 30). *What is Off Grid? | Living Off The Grid In Rural or Urban Areas Explained.* Off-Grid Expo. https://offgridexpo.com/what-is-off-grid/

Boni, T. (2008, October 10). *Living Off the Grid: How to Generate Your Own Electricity | Today's Homeowner.* Today's Homeowner. https://todayshomeowner.com/living-off-the-grid-generating-your-own-electricity/

Contributor, S. L. (2014, March 12). *Off The Grid Gardening Tips.* Survival Life. https://survivallife.com/off-grid-gardening-tips/

Eric. (2014, December 17). *Living Off The Grid: What Does It Mean?* Off Grid World. https://offgridworld.com/living-off-the-grid-what-does-it-mean/

eric. (2019, June 9). *The Realities of Living Off Grid.* Bubba On The Road. http://www.bubbaontheroad.com/2019/06/09/the-realities-of-living-off-grid/

finally time to think: pros and cons of living off the grid – cope with life. (n.d.). Retrieved from https://www.copewithlife.ca/introvert/finally-time-to-think-pros-and-cons-of-living-off-the-grid/

Food Preservation. (n.d.). New Life On A Homestead | Homesteading Blog. Retrieved from https://www.newlifeonahomestead.com/the-homestead-kitchen/food-preservation/

Food Storage and Preservation | Nutrition.gov. (n.d.). Www.Nutrition.Gov. https://www.nutrition.gov/topics/food-safety/safe-food-storage

Gardening Off the Grid: Mastering Water and Solar Systems. (2018, May 29). Hobby Farms. https://www.hobbyfarms.com/gardening-off-the-grid-irrigation-water-solar/

Getting Started With Off-Grid Water Systems For A More Self-Reliant Homestead • Insteading. (n.d.). Insteading. https://insteading.com/blog/off-grid-water-system/

H, S. (2015, September 7). *Underground Houses: The Ultimate In Off-Grid Living?* Off The Grid News. https://www.offthegridnews.com/extreme-survival/underground-houses-the-ultimate-in-off-grid-living/

Homestead Survival Site - How to Live Off The Grid in Comfort and Style. (n.d.). Homestead Survival Site. Retrieved from https://homesteadsurvivalsite.com/

How to Design an Off-Grid House - GreenBuildingAdvisor. (2017, June 2). GreenBuildingAdvisor; GreenBuildingAdvisor. https://www.greenbuildingadvisor.com/article/how-to-design-an-off-grid-house

How To Get Started Living Off Grid The Homesteading Hippy. (2014, November 3). The Homesteading Hippy. https://thehomesteadinghippy.com/off-grid-living/

How to Preserve Meat in the Wild Without Refrigeration. (n.d.). Know Prepare Survive. Retrieved from https://knowpreparesurvive.com/survival/how-to-preserve-meat-in-the-wild/

Hunter, J. (2016, October 19). *Off Grid Checklist: Become Self-Reliant with these Steps*. Primal Survivor. https://www.primalsurvivor.net/off-grid-checklist/

Instructables. (2014, June 30). *DIY OFF GRID SOLAR SYSTEM*. Instructables; Instructables. https://www.instructables.com/id/DIY-OFF-GRID-SOLAR-SYSTEM/

Living Off the Grid: Pros and Cons - GeekExtreme. (n.d.). Retrieved from https://www.geekextreme.com/design/living-off-grid-pros-cons-20519/

Living off the Grid: Starting From Scratch (Part 1). (2014, November 1). The Good Men Project. https://goodmenproject.com/featured-content/kt-living-off-the-grid-starting-from-scratch/

Living Off-Grid: What It's Actually Like • Insteading. (n.d.). Insteading. Retrieved from https://insteading.com/blog/living-off-the-grid/

Max, A. be ready". (2019, October 14). *How To Go Off-Grid Step-by-Step*. American Patriot Survivalist. https://americanpatriotsurvivalist.com/how-to-go-off-grid/

MorningChores - Build Your Self-Sufficient Life. (n.d.). MorningChores. Retrieved from https://morningchores.com

MOTHER EARTH NEWS | The Original Guide to Living Wisely. (n.d.). Mother Earth News. Retrieved from https://www.motherearthnews.com

Oetken, N. (2018, December). *10 Ways to Preserve Meat Without a Fridge or Freezer*. Urban Survival Site. https://urbansurvivalsite.com/ways-preserve-meat/

Off Grid World - How To Live Off The Grid. (n.d.). Offgridworld.com. Retrieved from https://offgridworld.com/

Off-Grid or Stand-Alone Renewable Energy Systems. (2020). Energy.Gov. https://www.energy.gov/energysaver/grid-or-stand-alone-renewable-energy-systems

Pacific Lutheran University BA, E., & Twitter, T. (n.d.). *Generating Off-Grid Power: The 4 Best Ways*. Treehugger. Retrieved from https://www.treehugger.com/generating-off-grid-power-the-four-best-ways-4858714

Pedersen, D. (2018, May 29). *Realities of off-grid living.* The Land. https://www.theland.com.au/story/5435370/realities-of-off-grid-living/

Primal Survivor. (2017, February 2). Primal Survivor. https://www.primalsurvivor.net/

Pros and Cons of Living Off Grid. (2019, September 1). Offgridmaker.com. https://offgridmaker.com/2019/09/01/pros-and-cons-of-living-off-grid/

R. Paul Singh, & H. Russell Cross. (2017). Meat processing - Livestock slaughter procedures. In *Encyclopædia Britannica.* https://www.britannica.com/technology/meat-processing/Livestock-slaughter-procedures

Raising Chickens Off the Grid ~ Without Heat Lamps or Lights. (2018, April 30). Practical Self Reliance. https://practicalselfreliance.com/chickens-without-electricity/

Schwartz, D. M. (n.d.). *Best Alternative Off Grid Toilets - No Septic!* Off Grid Permaculture. Retrieved from https://offgridpermaculture.com/Water_Systems/Best_Alternative_Off_Grid_Toilets___No_Septic_.html

The Reality of Living Off-Grid in a Caravan with Children in Central Portugal Over Winter. (2018, February 25). Topsy Turvy Tribe. https://topsyturvytribe.com/portugal/the-reality-of-living-off-grid-in-a-caravan-with-children-in-central-portugal-over-winter/

What Is an Earthship Home? Eco-Friendly Living and Zero Utility Bills. (2018, January 24). Real Estate News and Advice | Realtor.com®. https://www.realtor.com/advice/buy/earthship-home/

What is The Meaning of Living Off The Grid? (2020, January 15). An Off Grid Life. https://www.anoffgridlife.com/what-is-the-meaning-of-living-off-the-grid/

www.ingramcontent.com/pod-product-compliance
Lightning Source LLC
Chambersburg PA
CBHW050639190326
41458CB00008B/2336